Economic Strategy for Developing Nuclear Breeder Reactors

Economic Strategy for Developing Nuclear Breeder Reactors

Paul W. MacAvoy

The M. I. T. Press
Cambridge, Massachusetts, and London, England

Set in Times New Roman.
Printed by Publication Press, Inc., Baltimore, Maryland.
Bound in the United States of America
by The Maple Press Co., York, Pennsylvania.

SBN 262 13054 8 (hardcover)

Library of Congress catalog card number: 70–78097

For Matthew W. MacAvoy,
since he is a member of the
college graduating class of 1989

Preface

This review of research strategy was fostered by particular conditions of both "demand" and "supply," and it may be appropriate to mention them at the outset. The author, while a member of the staff of the President's Council of Economic Advisors in 1965 and 1966, found quite evident demands in both government and industry for a broader economic evaluation of the breeder reactor development program than that of the Atomic Energy Commission. The demands for benefit-cost analysis exist for every program under the large umbrella of the Bureau of the Budget's PPBS (Planning-Programming-Budgeting-Systems); but in this case they seemed more extensive because of the very large scale of proposed costs of the program and because little work was being done outside the agency that would carry out the program. There seemed to be a need for consideration of national economic benefits from the breeder, rather than concentrating attention on promised technical advances in efficiency of uranium usage for the reactor industry alone. A review of reactor research in the context of the United States economy might well justify a different total and composition of expenditures than those envisioned.

The study furnished here, in some part, meets the demand. But the author's interests have limited and focused the evaluation. Little attention has been paid to the question of whether the breeder reactor program should be eliminated: the findings in the last chapter lend no support to an affirmative answer. Rather, interest here has centered on alternative breeder programs — of longer or shorter duration, of smaller or larger size — offering greater or less benefits. Then the aim has been to find the research project mix offering the largest present value of forecast economic benefits over and above research costs. The program which seems to meet this standard greatly stresses carrying on duplicative research in separate companies at greater expense than the present Atomic Energy Commission program — both to increase the number of breeder reactor types so as to meet varied demands, and to increase the benefits from reduced industry concentration.

The conditions of supply have been lavish. Resources For the Future, Incorporated, has provided a grant which enabled me to attend courses in nuclear engineering, to consult with engineers throughout the country, and to devote other nonteaching hours to constructing this review. The R.F.F. grant, one of the more sizable that they have made, provided continued support beyond the original two-year period allowed. The sustained enthusiasm and interest in this project by R.F.F. made it all possible.

There have been many individual contributants, as well. The assistance of R. L. Schmalensee, J. A. Bewick, W. A. Stull, B. Unger, and Mrs. Eva Ewing at the Massachusetts Institute of Technology has been important in constructing the data file and computer programs in the four-equation model of the breeder and nonbreeder reactor markets in Chapter 3. A number of economists contributed detailed reviews of the March, 1968 version of this manuscript that led to substantive revision of Chapter 3 — including Robert Barlow (President's Office of Science and Technology), William Comanor (Stanford University), Orris Herfindahl (Resources For the Future), William R. Hughes (Boston College), Milton F. Searl (Office of Science and Technology), W. G. Shepherd (University of Michigan), Lester Thurow (M.I.T.), and John Vernon (Duke University). Perceptive and constructive reviews of other chapters were provided by Commissioner James Ramey of the Atomic Energy Commission, Gale Young and James Lane of Oak Ridge National Laboratories, Donald Marquis (M.I.T.), Daniel Taft (Bureau of the Budget), and a number of necessarily anonymous employees of the reactor manufacturing and research corporations. Mr. Phillip Sporn of New York City contributed many questions, most of which I was not, but should have been, able to answer. Any author's greatest windfall is a page-by-page critique by an energetic analyst doing research on the same issues; my windfall was such a critique from Richard Williamson of the Departments of Economics and Nuclear Engineering at North Carolina State University. All of these individuals are absolved of any responsibility for this study, but they are not free of my gratitude for their effort to improve the results.

Special note should be made here of the use of the M.I.T. economic simulation computer language. This TROLL language, as developed by Edwin Kuh and Mark Eisner of the Sloan School, made it possible to install and store in IBM 7094 time-sharing the four-equation model of regional nuclear capacity outlined in Chapter 3. This model was then used to simulate the 1985–2004 period for more than two dozen conceivable values of parameters in less than 30 minutes of computer time — work that required months of rough approximation for the March, 1968 draft of this manuscript. This new capability for "economic-sensitivity" analysis should be widely productive; I am grateful for the opportunity to be one of the first to appreciate its capacity and flexibility.

There is every reason to believe that the data for forecasting nuclear costs and benefits will improve in the next five years, so that a study of this nature published in 1975 could only be better than the present publication. This author hopes to do such a "second round." But comments, criticism, and encouragement have to be supplied by readers if the 1975 version is to be a substantial improvement; may the list of suppliers grow longer.

Boston, Massachusetts PAUL W. MACAVOY
November 1968

Contents

Economic Strategy for Developing Nuclear Breeder Reactors

1

The Questions in Future Reactor Research and Economic Approaches to Answers

With the purchase of nuclear reactors by most of the larger electricity generating companies in this country, the Federal Government reactor research program has gained some breathing space. The basic research on the two successful "light water" reactor types has been completed, with the collection of information on long-term operating performance the only continuing activity. Other research on other reactor types which promised to match or exceed the performance in the 1960's of "light water" reactors has been shut down or at least reduced to a much lower level.

The breathing space, however, promises to be short. The Atomic Energy Commission's conclusion is that "to achieve widespread commercial use of nuclear energy for the production of heat and electricity . . . will require the development of fast breeder reactors"[1] and such development implies reactor research on a larger scale than ever before. This work will involve an amount of capital that is at present unknown, but which could approach $2 billion of long-term commitments to new research on any single fast reactor type so as to attain "widespread commercial use" of that type in the mid-1980's. The research is to be financed in large part by the Atomic Energy Commission.

The new research on any one type of breeder reactor — termed a research "project"[2] — may promise the same results as other projects but achieves those results by solving different problems, or it may

[1] Milton Shaw (Director of Reactor Development, USAEC), "Fast Breeder Program in the United States," *London Conference on Fast Breeder Reactors* (17–19 May, 1966: British Nuclear Energy Society).

[2] A research "project" consists of the activities of a company or group of companies resulting in operation of a single reactor of a new type or technology.

promise quite different results in terms of the operating performance of prototype reactors. Choices have arisen and have to be made presently to achieve results for the 1980's and 1990's. They are on the number of projects — *how many* fast reactor types — and among the candidates for doing the work — *which* reactor types should be developed, by *which* companies, over *what time span.*

The reasons for Federal research in fast breeder reactors are diverse and somewhat inconsistent. These reactor types are to be pursued to reduce the demands on the known stock of uranium reserves, since they "breed" more fissionable plutonium from low-productive or partially depleted uranium than is fissioned per unit time. An alternative reason for the new research is that breeders increase the demands for the stockpiled plutonium by-product from present commercial reactors. An increase in reactor "efficiency" is sought, where this is measured in terms of the ratio of fissionable material produced to that consumed, and the ratio of electricity generating capacity to fuel inventory.[3] These are technical considerations; the economic consideration is toward developing a breeder power plant chosen by the electricity generating companies rather than fossil-fueled boilers or presently available nuclear facilities: "the objective of the fast breeder effort is the development of the economic power plant for the utility environment."[4] The shift of buyers' demands would follow from reducing uranium fuel costs to the generating companies below those of alternative plants. This may result from reducing demands for reserves or increasing "efficiency," but not necessarily both.

Whatever the nature of its prime objectives, the Atomic Energy Commission has already expressed preference for a reactor type. Three of a large number of breeder types have some claim for consideration as projects. Each has based its claim on having lower research costs: "the steam-cooled concept would make maximum use of water plant technology and the familiarity of the utilities with water-steam plants . . . the gas-cooled fast reactor concept would make use of technology which has been highly developed in the United Kingdom."[5] The Atomic Energy Commission's choice has not been dislodged by comparing costs with those of the other two: "It appears unlikely that the economic and technical comparisons will give any significant *advantages* to either the steam- or gas-cooled concept to the degree necessary to provide reasons for *changing* the high priority assigned to the sodium-cooled fast breeder reactor in the U.S."[6]

[3] These two ratios cannot both be increased at the same time, as will be seen.
[4] Shaw, "Fast Breeder Program in the United States," p. 3.
[5] *Ibid.*, p. 39.
[6] *Ibid.*, p. 39 (emphasis added).

The organization of projects into a research program is more of an economic matter than suggested in this choice, however. Two or three projects could be financed so that a number of companies would offer final demonstration reactors separately and "competitively." Alternatively, the projects could be consolidated and contracts let so that each company specialized in certain components or certain plant sizes and no two companies offered new equipment with the same specifications. The decision between the two possibilities — and among those within these extreme cases — depends upon the relative costs of research and relative benefits from the reactors actually built after the research is complete. That is, the questions to be asked in designing a Federal breeder research program include both "how many projects" and "how many companies." The answers are acknowledged to be in economic terms. But what terms these are, given uncertainty as to research costs and results from each project, remains to be specified in detail.

The economics of project selection is usually embodied in benefit-cost analysis. A quantitative benefit-cost evaluation of a proposed research and development project is a fairly straightforward matter. The costs of research are forecast for the project and year of occurrence and then are discounted to present values with the interest rate that seems appropriate. Benefits are the expenditures of consumers on goods and services derived from the research, plus any general environmental benefits, net of all costs of producing the goods and services. These benefits are discounted at the appropriate rate of interest as well. Where the present value of benefits exceeds that of costs, then the project merits consideration, and projects are chosen according to highest net present value.[7]

Has this set of standards been sufficient to explain the choice of particular research programs? Until now the answer has been in the negative. A number of successful research projects can be pointed to which did not show high or outstanding net present values before they were begun, and there have been others which had obviously high values but which were not undertaken. In the case of nuclear research, the submarine reactor must have been an example of the first (although not all of this project's *a priori* "benefits" were calculable from public information at the time the project was undertaken), and power reactors fueled with natural uranium and moderated with heavy water — as in the Canadian reactors — must have been an example of the second in the early 1950's. Certainly enough examples can be found

[7] This assumes that the calculated costs measure the opportunity costs of using resources in this research endeavor and that the "appropriate" interest charge is the social rate of interest for this project as compared to alternative projects.

to make it difficult to explain past allocation of research funds on benefit-cost principles.

There are good reasons for carrying on research other than those shown by cost-effectiveness calculations. The high "net present value" may not justify the risks that characterize this research. For one thing, the expected net present value (or benefits minus costs) of a nuclear research project might be shown to be as high as that for investment in a new bridge, but the probability that these benefits and costs will be realized in the first case may be only 1/10 the probability in the second. The most likely reaction is to defer the nuclear work in favor of building the bridge when there is competition between the two for project funds, even though the expected returns are the same. After all, most Government officials and corporate officers find it difficult to justify to the final claimant — the voter or stockholder — wagering all on the long shot project only because it is the long shot.

Another good reason for going beyond a benefit-cost calculation for a research project is that, in most cases, no single estimate can be made of project results which is more likely to be realized than a number of other estimates. In other words, results in the far-distant future that are demanded by buyers with changing alternatives offer a range of benefits net of costs. The benefit-cost framework applies to each estimate for each project, but does not provide a single "net present value" to be compared with those of other projects.

Can economic analysis assist in providing a framework beyond the benefit-cost framework to guide decision-making on such uncertain research projects? The answer here is that it can help to formulate a rational reaction pattern to risk which accompanies research and development.

Economic Evaluation of a Research Project without Explicit Consideration of Risk

Consider the quantity demanded Q for the product of a research project in one year to be shown by $Q = f(P, Y)$ for prices P for this and other products and incomes Y of consumers. The gross benefits from successful research are equal to the total amount which consumers are willing to pay rather than continue in a world without the product. These are the sum of two revenue streams shown in Figure 1.1: the curve D indicates the quantity $Q = f(P, Y)$, where the first area P_0Q_0 indicates total payments for the amount Q_0 at the going price P_0, and the second area P_1DP_0 indicates the amount the consumers gain from having to pay only the marginal price for all intra-marginal quantities. The second area, the graphical illustration of "consumers'

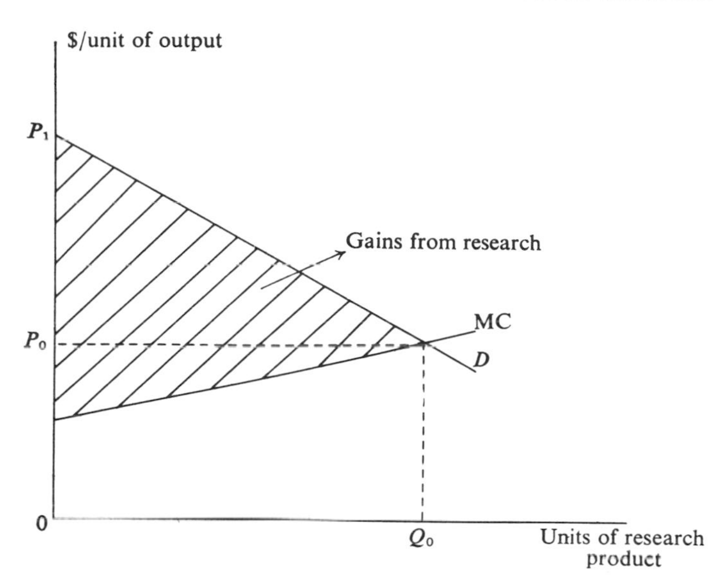

FIGURE 1.1 The market for the product of a research project

surplus," is counted in the results from a research project but not a typical plant and equipment investment project because research implies all or zero output — and the gains from all intra-marginal units — while the typical plant investment plan is only for marginal additions to output.

The net gains to society from this research, as contrasted to the gross benefits to consumers, are equal to the present value of these annual gross benefits, minus the total costs of the final product and, of course, the costs of the research itself. Total production costs each year are $C = f(Q)$ and are shown as the area below the marginal cost curve MC($=dC/dQ$) up to Q_0 for that output demanded at price P_0. The benefits exclusive of production costs are then shown as the shaded area in Figure 1.1. The sum of the same calculations for all years can be termed "total production benefits." Net research benefits are equal to this amount minus the costs of research; or, in present value terms, the "net social benefits" from research are equal to the present value of production benefits minus the present value of research costs.

These gains from the project can be made small or large by the policies of the organizations producing the goods resulting from the research. For the largest possible production benefits, the outputs of these organizations should total to Q_0 because the value of the last unit is equal to the costs of resources to produce this last unit at that rate of output. But if the decision-making organization is a single private company, then Q_0 is not the decision-maker's most profitable output. The dollar equivalent to the area under D between P_1 and P_0 accrues as surplus to final consumers; the remaining shaded area between P_0 and MC is the surplus accumulated by the company. To maximize private company profits, output should be less than Q_0 because the smaller output adds more to the difference between price and marginal costs on all units provided than lost in the low price-cost margin on the output foregone [the most profitable company output is that for which profits $\{P \cdot Q - C(Q)\}$ are a maximum, rather than that for which social benefits $\{\int_0^Q P\,dQ - C(Q)\}$ are a maximum, and the first is a smaller amount].[8] Restriction of output requires monopoly or oligopoly power — power to change the total amount of the product offered for sale. The restriction, if practiced, costs the consumers more than the producer gains, by an amount shown by the price-marginal cost difference for all of the units by which output is reduced.[9] The restriction will be practiced if the research results are limited to few companies. Then the policies for distributing the results have implications for the size of "net present value" on the project.

This calculation of social benefits and costs on research, even though straightforward, differs in two ways from the usual analysis. The first disparity is in the treatment of consumers' surplus. If the results of research include a new product, rather than only a cheaper way of producing an existing product, then the economic argument is that the value of research exceeds potential expenditures on the product by an amount shown as the shaded area P_1DP_0 under the demand curve. There is "consumers' surplus" for new products, and since the research generates the entire market — there is zero output if the project is not put into effect, but positive output from taking on the project — it can claim all of the surplus. The amount claimed may be important: in Figure 1.1, as an illustration, this surplus is not only a source of gross benefits, but most of the net benefits. The second disparity here is in not accepting the market price of the research output as the basis for finding net social benefits. The price is one of the products of the research program of the Government — since the program determines the structure of the market for the final output from any research

[8] The first-order conditions for max $\{P \cdot Q - C\}$ are $P + Q\dfrac{dP}{dQ} - \dfrac{dC}{dQ} = 0$ while those for max $\{\int_0^Q P\,dQ - C\}$ are $P - \dfrac{\partial C}{\partial Q} = 0$. The first implies a higher price and thus a lower output since $dP/dQ < 0$ (the lower the price the larger the quantity demanded).

[9] A single-buyer policy would have similar effects in the opposite direction: with the cost curve shown and the market rule that all units are sold at the same price, max $\{\int_0^Q P\,dQ - (dC/dQ)Q\}$ requires that $Q < Q_0$.

project. Estimates based only on prices given by firms in the most exclusive project may show less than full net benefits.

Economic Decision Rules under Risk and Uncertainty

The identifying characteristics of research are that expenditures on a project to achieve given results are not certain, nor are consumers' reactions to the results certain. A single benefit-cost calculation does not show the effects of these attributes, but multiple calculations can be made that do. The economics of investment sets out rules for evaluating projects, given that uncertainty can be shown by the variation among possible net present values on a single project.

The fruits of research include new products, the demand for which is "unknown," in the sense that the location of curve D cannot be fixed but is indicative of one probable price-quantity relationship. There is some probability that P_0 will clear the market of Q_0; there are other finite probabilities of both higher and lower prices for Q_0. The cost function is also unknown; marginal costs may be MC with a certain probability, but twice as great or half as great with other probabilities. Then the shaded area in Figure 1.1 equals the annual gross benefits from research which occur with a probability given by the product of the probabilities of P_0 and MC. Other shaded areas, from the occurrence of other levels of price and marginal costs, have other probabilities. The calculation of benefits, for each combination of conceivable price and costs, results in a frequency distribution of production benefits from research.

This distribution is for one year of benefits from adoption of the research results. The distribution for total production benefits from the project is derived from the sums of the present values of each year's benefits. These sums are calculated by adding together the present discounted values of the annual benefits from one price and cost result, and then separately for another result. Then the distribution is found from observing the frequency of occurrence of a particular sum (given the underlying probabilities of P and MC).

The costs that will be incurred in completing the research are also unknown. The engineering and construction tasks might be specified in some detail, but no one can specify exactly how many dollars will be needed to complete them. At best, a frequency distribution of research expenses can be found by subjective assessments of the probability of costs of each task, and then of the sum of the costs of each task. The two distributions — of production benefits and research costs — generate a frequency distribution of net present values.

Each potential project has such a distribution of its own, conceptually at least. Each project's worth can be described, as a first approximation, by the characteristics of this distribution: mean, variance, and skewness. Some projects can be said to be better than others on the basis of these dimensions: of projects with the same mean net present values, the project with the smallest variance and downward skewness is best.[10] There are two reasons for this. First, it is assumed that the results of research go to final consumers who prefer more to less, but additional satisfactions gained from another unit are always less than on the previous unit. Given this assumption, the project with larger variance offers the probability of small additional satisfactions gained from realizing high present values, but of large satisfactions lost from the low present values. The average of satisfactions gained is lower for this project than for one with the same average but lower variance. This project should be rejected to maximize consumer satisfactions (whether satisfactions are ever measured or not). Second, the probability of zero or negative net present values should be the dominant factor in choice. A project with the chance of default should be rejected for one without that chance.

Let us consider two research projects, each of which is thought capable of providing a marketable product "*Q*" at the end of a five-year development period. It may well be that their most likely net present values are the same, so that both expect to cost $5 million in the one-year research period, and expect to provide a stream of revenues with mean present value of $6 million from five years of product sales thereafter. But enough is known of the research problems in the first project to say that the chances are equally as great that total research costs will be $500,000 greater or less than expected. The second project is straightforward; indeed, it hardly qualifies as "research" since the expected costs have 1.0 probability.[11] The distribution of present values from the first project centers at $1 million, but would have values at $1.5 million and $0.5 million. The distribution in the second project would consist of one point at $1 million. The second project has to be the better avenue of research because it offers reduced variance.[12]

[10] Cf. H. Markowitz, *Portfolio Selection: Efficient Diversification of Investments* (New York: Wiley, 1959); and, for lucid examples of this rule in the choice of capital budgeting projects, cf. G. D. Quirin, *The Capital Expenditure Decision* (Homewood, Ill.: Irwin, 1967), Chapters 10, 11.

[11] This is not to suggest that estimation of probabilities is a simple and straightforward task. As will be seen in the example of the breeder reactor projects, there are formidable problems in analyzing research plans so as to include the possibility of alternative results. The formulators of projects do not provide for variations from their plans. But some estimate of probabilities — even of relative magnitude, as above — is better than none, because consistent decision rules can then be applied.

[12] Given that less variance — less chance of departure from design — is better than more. Such an assumption — of declining marginal satisfactions — seems to conform to observed behavior in all except the most special cases. Cf. Quirin, *The Capital Expenditure Decision*, p. 203.

The choice made in another circumstance has unusual results. The decision-maker has available to him the plans for four projects, all designed to develop the same product; the projects all have the same forecast costs for developing this final consumers' good, but each takes a different technical route — solving a different engineering problem — to the final objective. One research plan is to undertake all four projects for $200,000 each and then, after some first-round results, choose the most promising to complete the task (with a further expense of $200,000). A second plan calls for putting $1 million into a single project from the beginning. After some review of the technical problems in each of these projects, the decision-maker has concluded quite subjectively that the probability of success of any one of the four projects is 1/4 and of failure is 3/4 — except when $1 million are assigned to one single project, so that the chance of success is 1/2.[13] Then successful research in the first plan has a probability of 7/10 [the probability of four failures is $(3/4)^4$, which is one minus the probability of at least one success], but only a probability of 1/2 in the second plan. The dominant program is the first, which undertakes all four projects at low levels until one of the four proves out.[14]

The choice of projects on the basis of obvious characteristics of the "net present value" distribution may not be sufficient to discriminate among all alternatives. Some one project might have high mean net present value but also extremely high variance in values. It promises more, but with more risk.

In this case the Government can serve well as decision-maker. As the holder of a large portfolio of research projects — some for consumption of defense goods, others in collaboration with private industry for the development of final consumers' goods — the Federal Government can add part or all of the project to those it has under way. This participation reduces the risk for the companies included, so that it is comparable to that in other private projects (the Federal funds change the distribution of private net present values on this project so as to make average, variance, and skewness comparable to those for alternative private projects). From its own point of view, the Government adds the new project to its portfolio so as to change the average and variance of the net value of that portfolio. As the largest holder of risky projects, this

[13] In each instance, complete failure is assumed to be the result of "lack of success." There is no partially successful project, so that "lack of success" does have a probability of 3/4 for each project.

[14] This case would seem to be an example following the appeal of B. H. Klein for duplicative projects as a cost-saving method of solving research problems. Cf. B. H. Klein, "The Decision-Making Problem in Development," in *The Rate and Direction of Inventive Activity: Economic and Social Factors*, National Bureau of Economic Research Report (Princeton, N. J.: Princeton University Press, 1962).

change is almost always not very great; the result is to reduce risk to particular industries by balancing it with that from other projects.[15] If the change in value of the Government's overall research program is great, then less participation is called for and the project may be foreclosed.[16]

Government rules for deciding in all circumstances whether to start a project, to continue it, or to up-grade it, comprise a "strategy." The economic strategy calls for calculating the net present value from each possible research result from a single project and then setting out the (subjective) frequency distribution of these net values. Then the frequency distributions of project mixes are set out. Strategy consists of choosing the project or project mix with highest average, lowest variance, and lowest negative skewness in the frequency distribution of net present values. This single rule produces most of the reactions to risk, or most of the correct economic strategies.

Strategies for Development of the Fast Breeder Reactor

The breeder reactor can be developed by recourse to one of a number of alternative programs, each with a different projection of required expenditures of time and money, and each with a distinctive promise of final benefits. All programs are risky. The development problems may not be solved until twice as much is spent as originally thought necessary. The benefits may be different from those expected because of surprising conditions in fossil fuel or uranium markets. But the amount of both research expenditure and the resulting benefits will be billions of dollars. The questions are whether application of economic strategies — at this time, in a preliminary way — might not promise to reduce *projected expenditures* and increase *projected benefits* so as to make an appreciable difference in the amount of billions of dollars.

The Research and Development Problems

The reactor fuel core, consisting of uranium, plutonium, and coolant encased in various metals, seems less complicated in the fast breeder reactor than in the (present-day) thermal reactor because it does not

[15] This is by analogy. Here the Government is the market for risky ventures, and has the same effect as that from individuals balancing their private investment portfolios, so as to clear the securities markets. For that effect, see E. Fama, "Risk, Return and Equilibrium," *Journal of Finance*, Vol. XXIII, No. 1 (1968), pp. 29–40.

[16] That is, a higher risk premium has to be used in the discount rate applied to find net present values.

contain a moderator. But it has much more complex behavioral patterns. The neutrons released in the fission of Plutonium Pu 239 possess high kinetic energy and retain this energy in the absence of the moderator when scattering through the core. Those neutrons in the "fast" environment colliding with natural uranium (U^{238}) — a cheap "blanket" material which does not fission — cause the conversion of large relative amounts of U^{238} to plutonium, the original fissionable material.[17]

This reaction varies widely fron one second to the next. The reactivity ρ — the percentage rate of birth of neutrons in fission in excess of those holding the neutron population constant — changes with coolant temperature and the stochastic changes in temperature are an order of magnitude quicker and larger in a fast reactor. Then heat energy production more probably will exceed the maximum allowed by metal temperature limits, and the core will "melt down."

The fast reactor contains all of the usual subsystems — fuel rods assembled into a core, heat exchangers, turbine generator leading to a condenser. But the core is extremely compact because of the absence of a moderator and because of attempts to achieve the higher power densities (kilowatts per gram of fuel) that make up for the economic burden of high initial plutonium content. The heat absorption capabilities of the coolant have to be of the highest order for the same reason. Thus the core is subject to extreme temperatures in the event of accidental breaks in the system. Tubing and structures have to be developed which are "break proof" to minimize such occurrences. Also, the fuel element in the fast breeder reactor is especially difficult to construct.

The fuel rods are of extremely small diameter, to achieve the high heat transfer capabilities required. They require more of materials and design than now achieved in the construction of thermal (present day) reactor fuel cores.[18]

The fast breeder, difficult to construct, is designed to operate at the extremes of metallic tolerability to heat, and in a fashion difficult to control. The surprise is that there are at least three candidate systems.[19] The *liquid metal fast breeder* uses liquid sodium as the coolant to take heat from the core through multiple heat exchangers to steam turbine generators. There are advantages in doing so — sodium has

[17] When the plutonium produced exceeds the volume originally inserted in the core per unit time, then the apparatus is said to "breed."

[18] Cf. T. J. Thompson and J. G. Beckerley, *The Technology of Nuclear Reactor Safety*, Vol. 1 (Cambridge, Mass.: The M.I.T. Press, 1964), p. 158.

[19] There is a host of further candidates other than those discussed here. But these three are blessed with designs for 1,000 MW_e systems, working mock-ups, experimental components — even working prototypes, in one case.

excellent heat absorption capabilities from 200°F to 1,600°F at atmospheric pressure — but sodium at these temperatures welds joints and bearings, explodes on contact with air and water, and is so absorptive that thermal shock is transferred quickly from one spot in the core to all parts of the system. The *steam-cooled fast breeder* passes low-temperature steam from a pre-heat boiler through the fuel core to produce high-temperature steam which goes directly to the turbine generator. This simple system avoids heat exchangers and any special enclosures to isolate the coolant, but it delivers corrosive radioactive steam to the turbine generator. The closed cycle also raises problems of maintaining steam under pressure in the core in the event of a break in the system. The *gas-cooled fast breeder* combines the strong and weak aspects of the other candidates. Helium is transmitted through the core under high pressure to a set of heat exchangers where steam is then generated. The safety of this coolant matches that of steam, but the high pressure of the system is as much a disadvantage in both systems. The compartmentilization of a gas fast breeder is similar to that in the liquid sodium breeder, as is the complexity of heat exchanger mechanisms.

Designs have been completed of reactors in each of these three systems which are to produce electric power for less than four mills per kilowatt hour. In the case of the liquid metal system, prototypes have been constructed in the United States (the Enrico Fermi Plant), Great Britain (Dounreay), and the U.S.S.R. (the BN series). But these reactors have not provided continuous power capacity to commercial standards. A meltdown of a core section of the Enrico Fermi plant after intolerable core temperatures rendered that plant inoperable through a good part of 1967, for example; the others have been carefully operated to carry out controlled irradiation of fuel rods, and incidentally to produce steam. There are no comparable prototypes for gas- and steam-cooled systems. The fast reactor remains to be developed.

The problems to be solved before the construction of the first demonstration fast breeder reactor are formidable. But they are not impossible. There is some expenditure on small-scale plants that will "prove" systems so as to make it possible to proceed with large-scale demonstration plants. This expenditure may be exceedingly large, but research results can be purchased.

These costs of research are amenable to forecasting. Any number of forecasts is possible and at least three forecasting techniques come to mind that would provide different estimates of total expenditures from the initiation of a program in the early 1970's to termination of the research portion of the large-scale demonstration reactor in the mid-1980's. The first method is to extrapolate the behavior in past reactor development: given the relation between research results and outlay for the two completed programs, the additions to outlay to complete the fast reactor program can be forecast. This is done by taking the multiple of expenditures required to move from the present results to terminal full success where that multiple is given by the ratio of terminal/initial expenditures on the past programs at roughly the same stage of development. The second technique is to ask a sample of knowledgeable nuclear engineers what the costs of a fast breeder program will be, given specific targets to reach by 1985. The third technique is to do an economic and engineering review of the technical problems in each subsystem of each breeder reactor project and then set a range of "price tags" for the solution of each of them. The sum of any one set of price tags equals one estimate of the total costs of a fast breeder reactor project.

The estimates of costs from these three calculation procedures could well differ by substantial amounts. Before the fact, the best procedure might well be to combine the three by giving subjective weights to each. The first could not be weighted heavily, given the sketchy and preliminary documentation of interim results from past fast breeder research. The combination has to put considerable weight on the opinions of experts on the nature of problems and on the detailed analysis of costs of solving these problems.

Forecast Benefits from Successful Introduction of Fast Breeder Reactors

After demonstration plants show that the breeder will produce thermal energy for sustained periods, then electricity generating companies will seek contracts for these breeder plants. The contracts will require manufacturers to build nuclear plants with capacity to produce rated electrical megawatts on designated load factors. The number of megawatts will depend upon the prices at which breeders are offered, and their expected fuel cycle costs as compared to thermal nuclear plants and fossil fuel electricity generating facilities. This demand-price relation is shown in Figure 1.2, in two steps. The benefits from the breeder are the producers' and consumers' surpluses given this demand.

There is first the demand by generating companies of thermal or nonbreeder reactors in the market for energy-transforming equipment. This is shown in Figure 1.2, where D_T is the demand curve for new thermal reactor capacity in some properly defined regional market as a function of the dollars per kilowatt of charges made by the equipment manufacturing and constructing companies. The curve has a negative slope: the lower the nuclear charges, given a schedule of charges for fossil fuel plants, the larger the nuclear megawatt capacity demanded in that power region. Institutional conditions — such as national

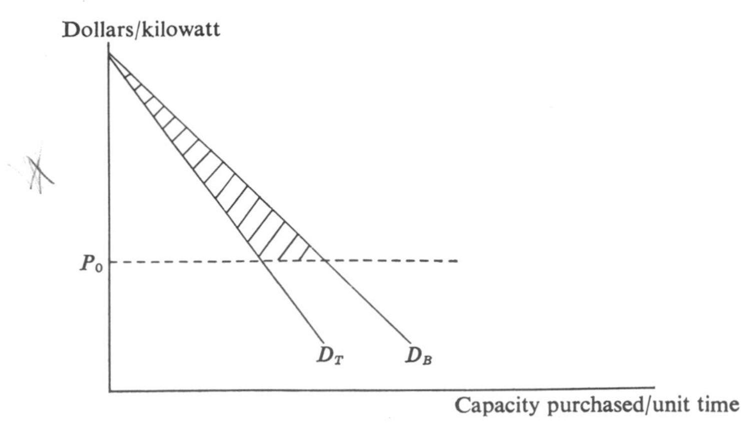

FIGURE 1.2 The demand for breeder capacity

regulation of the sales of electricity produced by reactors and fossil fuel plants which favors reactors — may make the demand function less elastic than otherwise; but the curve D_T should have a negative slope under all conditions. The location of the curve depends upon fuel costs and the costs of alternative sources of supply of generating capacity, such as that provided by fossil fuel equipment manufacturers. The general presumption is that, the lower the reactor fuel costs and the higher the fossil fuel and capital costs, the farther upward and to the right is the D_T demand curve.

The second step is to add the net demand for breeders. This is shown as D_B in Figure 1.2 — as the shift outwards in D_T caused by introduction of the further special qualities of the breeder. The important special quality is reduced fuel cycle costs which result from this reactor "breeding its own." Whatever the supply conditions in the nuclear fuel markets and existing D_T demand for reactors, reduced fuel costs from breeding will shift out the demand for nuclear capacity. This difference between thermal demand D_T and breeder demand[20] D_B is the measurable gain in demand for reactors from having the new technology available. The dollar equivalent to the shift of demand — shown in Figure 1.3 as the cross-hatched area — will be one of the fo ecast benefits from the new reactor type.

This accounts only for the amount the consumers would pay, over and above the quoted contract price P_0, for the new reactor rather than

[20] It should be noted that D_T and D_B are mutually exclusive — both show the *total* demand for reactors, the first when only nonbreeders are available and the second when both breeders and nonbreeders are available. The second, then, either shows the demand for breeders with no demand for nonbreeders or the demand for that combination of the two systems offering *minimum* full cycle costs.

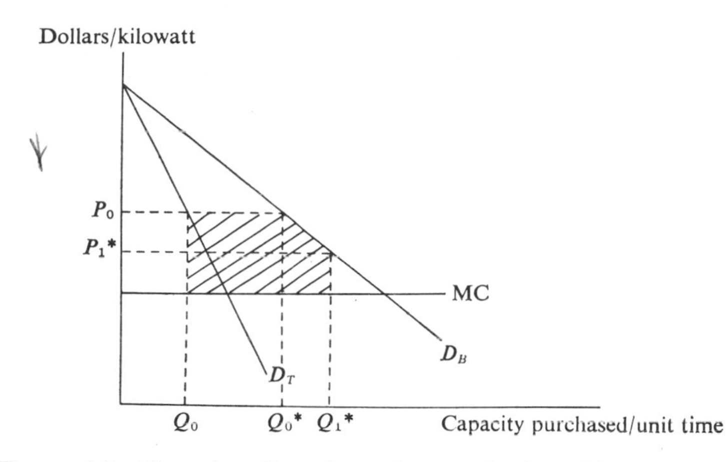

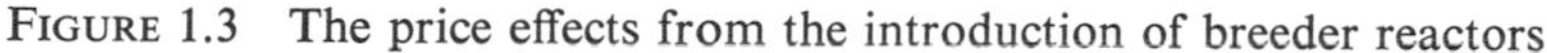
FIGURE 1.3 The price effects from the introduction of breeder reactors

go without these results from new research. The reactor producers gain as well, in keeping with the increased demand and the prevailing margin of price over component and construction costs. This source of benefits is shown in Figure 1.3. The increased demand at P_0 is $Q_0^* - Q_0$, and the gains in profits are the cross-hatched area between P_0 and marginal production costs MC on that additional demand.

There are other benefits. The introduction of the breeder reactor could reduce prices per kilowatt of installed capacity for all additional reactors. If *new firms* produce this new type of reactor, then the results could be the same as if there were more firms added to the existing market for nonbreeder nuclear power plants. The price might not be reduced, because not enough companies exist even after entry to allow "competition to break out." But the possibility exists for noncollusive pricing, in which case the adding of new firms reduces price-cost margins for reactor manufacturers in inverse relation to the number of firms. As price is reduced from P_0 to P_1^*, in Figure 1.3, installed capacity is increased from Q_0^* to Q_1^* with a consequent increase in net benefits shown by the cross-hatched area between Q_0^* and Q_1^* under D_B and above the marginal cost of newly installed capacity MC.[21]

The forecast is for economic benefits from the breeder from three sources. The first is consumers' surplus in the increased demand resulting from lower fuel cycle costs associated with breeding, and the second

[21] The costs of capacity may also be substantially decreased by the introduction of breeders. After research is complete, the results may well exceed forecasts for the performance of the nuclear reactor construction companies, and the marginal costs of installed capacity may be much less for breeders than now being forecast for thermal reactors. Then the gains to the economy are shown by the cost-difference per megawatt of new capacity times the amount of capacity Q_1^*. This cost saving is speculative; there is no basis for forecasting that it will be realized.

is producer's surplus on this same increased demand. The third is producers' and consumers' surpluses from the price reductions following the entry of independent firms producing the new reactor.

All three sources are quite suspect. The surpluses may never appear, because other characteristics of the breeder inhibit demand so that the differential between alternative demands D_B and D_T does not appear. For one, these are going to be new and untried reactors even in the 1980's, and the utility companies could decide against new technologies, given that the more established reactor types are operating consistently. Prices may not be reduced because the firms producing breeders are those already supplying thermal reactors. Most important of all, the benefits to producers and consumers of reactors may not be equal to total net social benefits. Those affected are particular producers and consumers whose demands may be as much set by electricity sales regulation and market imperfections as by the final benefits to the ultimate consumer of electricity.

Costs, Benefits, and Nuclear Development Strategy

Even though forecasts of the effects of breeders on the economy are most tentative, there is general agreement that the benefits could be very large. The demand forecasts generally indicate that from 500,000 to 800,000 MW of nuclear capacity will be installed between the years 1980 and 2000. If the savings in fuel cycle costs following the introduction of breeder reactors added only 10 per cent to nuclear demand at the expense of fossil-fueled generator demand, then the first and second sources of benefits could not accrue to less than $1 billion in this 20-year period.[22]

A price reduction on all new capacity, whether thermal or breeder, of $5 per kilowatt would be modest in the absence of perfect collusion among reactor producers, but would generate $4 billion of market-wide price reduction and (at least) one-fifth this much consumers' surplus from additional demand for capacity. There is little basis for arguing that the benefits after 1980 will be less than $2 billion of expenditure on research before 1980. The question is whether these social gains are great enough to justify a single large-scale project to develop a fast breeder reactor.

The choice of whether or not to carry out research and then develop a fast breeder reactor depends on more detailed forecasts of economic

[22] The most severe assumptions are that 20 per cent of prices of $100 per kilowatt are producers' surplus and there is no consumer's surplus. Then, on 500,000 to 800,000 MW of newly demanded capacity, the benefits are from $1 billion to $1.6 billion.

costs and benefits from such a project. The procedure for obtaining these forecasts is as follows. The liquid metal fast breeder reactor (LMFBR) project requires uncertain outlays each year from 1970 to 1985; the best present forecast of these outlays is C_i and their present value is $\sum_{i=1}^{15} C_i/(1 + r)^i$ where r is the appropriate social rate of discount of this type of research and development expenditure. This project promises results, as well, equal to $\sum_{j=15}^{n} B_j/(1 + r)^j$ in dollars equivalent to the benefits shown from the shaded areas in Figures 1.2 and 1.3 in each of those years in which new capacity is installed. The net present value $\{\sum B_j/(1 + r)^j - \sum C_i(1 + r)^i\}$ from the first 15 years of outlay and the succeeding years of economic gains may or may not be large enough to justify taking on the project.

The liquid metal project can very well require costs other than C_i to obtain the sought-after research results, and can produce a reactor generating more or less than the expected demand response by utility companies. There is a finite probability that research outlays will exceed C_i and be equal to C_i^* in any year and that the gains obtained therefrom will fall short of B_j and actually be equal to B_j^* in succeeding years. Then net present value would be $\sum B_j^*/(1 + r)^j - \sum C_i^*/(1 + r)^i$; before the outlays are made, this value, termed V^*, can be said to have a probability of occurrence P_r^*. There are other values of V with other probabilities of occurrence; together they generate a frequency distribution of net present values.

The conceivable cost and benefit results from building demonstration models of other reactor types can also be shown as frequency distributions of net present values. Subjective judgments have to be made of the probability of occurrence of each level of costs and accompanying level of benefits. These are arrayed with the resulting net present values. Preliminary to determination of strategy, there should be rough frequency distributions of V, at least, for the most likely breeder research projects.

The economics of strategy calls for a comparison of these distributions. If one of them has both the highest expected present value and the lowest dispersion from this value, then this is the preferred project. The question is then whether the expected value is higher than that for alternative research and development projects with comparable dispersion, where these alternatives are competing for government funds or are in the private sector competing for the research funds of companies. If the net present values on the most favored breeder project are higher than these alternative returns on research, then this single project should be undertaken at least during the initial stages.

The choice of the favored project, and of this project over alternative research in other industries, is more difficult when the higher average present value is accompanied by more variation among possible present values than is found on other projects. Then the choice is between a project with a higher average net present value but more risk (as measured by the variance, for example) and a second project with lower average value and less risk. But even here the choice may not be difficult because the higher variance may follow from greater positive skewness in the result: the variation from average net value is from possible higher net values in great part, and this variation favors the higher average, higher variance project.

Economics of strategy does not end with choice of a single project, because it may be possible to show that the addition of a second project also provides net present values greater than those on alternative expenditures of the funds. The successful completion of a gas breeder reactor program, for example, would require separate research expenditures beyond those called for in a liquid metal program (if the liquid metal program were chosen first). The benefits from the gas-cooled breeder would not be shown by the cross-hatched area in Figure 1.2 because the first breeder program would already have accounted for these gains (although much lower fuel cycle costs from a second program might shift D_B farther upward and to the right and the additional cross-hatched area resulting from the shift would have to be credited to the second program). But there could be price reductions as a result of the added program because there could be at least one additional *new firm* producing nuclear reactors as a result of specialization of an outside firm in gas-cooled reactor research. Then the cross-hatched area between D_B and MC over the capacity range Q_0^* to Q_1^* in Figure 1.3 should be credited to the new program (with P_1^* equal to the price schedule after the second breeder has been introduced). The net present value from the price reductions, even with "wasteful" or "duplicative" research, may be greater than those from alternative projects.

The best program could be a combination of a number of projects to build different types of breeder reactors. A plan first to solve basic core physics problems in the control of fission reactions without a moderator and then to construct experimental systems using gas, liquid metal, and steam could cost less than the sum of separate gas, liquid metal, and steam reactor projects. This is because the one program allows more work to be done in series with less overlap of research in separate technologies. All of the benefits of substantial price reductions from the addition of new firms could be captured by having independent firms bear the responsibility of carrying out the experimental systems research. Separate companies with separate breeder types from the consolidated program might well produce the maximum possible increase in net present values from research.

This is schematic economic strategy to be introduced at the first stages of project evaluation. As data are collected on research performance, the same schematic strategy applies later, however. Further information on the cost of research of various projects may change the rankings of projects by changing the averages and variances of net present values. When new information is sufficient to allow recalculation, then the strategy should be reapplied: choose the project with the best frequency distribution of net present values, and then add projects with distributions better than those of alternative research programs in other industries. Finally, choose a consolidated project made up of separate breeder programs (with research merged where it is identical in the separate programs) if the distribution of net present values for this project outranks that of the best single or separate project.

The first application of strategy on the breeder is set out in the chapters that follow. The range of costs of research on two reactor types is estimated. First crude indications of benefits are made from plausible demand and reactor production cost conditions and put against the range of research costs to generate frequency distributions of net present values for each type. Combinations of projects are evaluated in the same manner. The estimates of V, given the strategy, set out a plan for the near future.

2

The Costs of Fast Breeder Research and Development Programs

Research groups in the national laboratories and private corporation working on the development of fast breeder reactors all have the sam loosely defined goal — a breeder design that convinces electricit generating companies to construct these reactors. The means fo convincing them is to offer lower fuel cycle and capital prices. Th paths to this goal in engineering problem-solving diverge for gas cooled, steam-cooled, and liquid metal-cooled reactor research, but th general research techniques are quite similar. Each requires some wor in basic fuel core physics and the development of a "first-of-a-kind" system to generate steam for electricity. Improvements in the com ponents and in the routing systems of this "first prototype" are buil into a second stage reactor; costs of power are measured at this stage and, if there is promise of greater reductions, a third round will b constructed as a large-scale demonstration reactor. The demonstratio reactor will do what its name suggests — show the buyers that particular fast breeder type can produce steam consistently withou fuel cost disadvantages.

The technical hurdles which must be overcome to reach this objectiv are formidable. Physics research must be completed at the most basi levels before the development of fuel elements capable of the fissio burn-ups called for in the breeder design studies. Applied research towar finding materials that can be used under higher temperature operatin conditions — including conditions beyond those considered "normal" in the reactor design — has to be done in laboratories with small-scal models of fission processes and then in experimental reactors. Workin components have to be developed, for example, for the coolant systen in place of high-cost experimental components. Last and perhaps mos difficult of all, reactors have to be built that operate. The whole i greater than the parts put together, since successful integration of a parts of the apparatus into a workable system requires control of bot normal and abnormal interactions of the components. In the history of reactors, the development of low-cost components proved to be only a small item in the task of putting together a workable system; there seems little reason to believe that this will not be the case in the more complex systems required for fast breeding.

Solutions in each of these problem areas require extensive research and development expenditures. The review of the program that follows is for the purpose of forecasting these expenditures, to show their expected levels for a normal development path and also those levels given either research "breakthroughs" or research "setbacks." These three levels of costs are polar cases, because at least two reflect the extremes of the probable results; but they are also more than just contrasting cases, given that research results occur within each of *three categories of events*. The *first category* is made up of windfalls — easy or inexpensive solutions to major problems of materials development, mostly from work being done in other fields. The assumption is that, when the most probable of these breakthroughs occur, then the "low estimate" of costs is realized. The *second category* includes all those occurrences required to keep research and development on the time and cost schedule now laid out for any particular project for the period 1965–1985. Costs of research in this category have been estimated by the corporations and national laboratories (in some cases, extrapolated beyond the present five- or ten-year planning period) so as to obtain the funds from the Atomic Energy Commission to carry out the projects; otherwise, costs have been estimated here as those consistent with the proposed research path and then tested against the reactions of a number of research project managers for comparability with what they would submit. The *third category* is made up of events that can only be called serious failures in development of important components. Not all parts of a project are assumed to fail, but those viewed with the most suspicion are treated as if the fears were well founded. Costs in this case are those from projects which finally reach the design goal, but only after substantial additions of program inputs beyond those forecast.

The estimates are for those events either widely expected or widely feared, given the evaluations of the fast breeder appearing in the nuclear engineering literature. But they are all constructed, or put together from widely scattered sources, here for the first time. They stand only on the worth of the procedures by which they were made.[1]

[1] The estimates are complete for the liquid metal and gas-cooled reactor types. They are incomplete for the steam-cooled reactor because of early, extremely pessimistic results in both core performance and components development. Evaluation of steam-cooled reactors will have to await further documentation of initial engineering work.

The Liquid Metal Fast Breeder Reactor: Fuel Element Development

Reactor designers are fond of saying, "Provide me with a fuel element and I'll build you an efficient reactor." The first development in the LMFBR is a fuel rod that lasts longer and makes possible more complete burn-up of the plutonium than is now being experienced.

Fuel cycle costs are extremely sensitive both to the extent of the "burn-up" of the original inventory of plutonium, and to the ability to produce thermal energy from fission of a unit weight of this inventory. The target burn-up for LMFBR fuel has been set at 100,000 megawatt-days per metric ton (MWD/T); if only 50 per cent of this rating is achieved, 0.5 mills per kilowatt-hour (mills/kWh) are added to the fuel cycle cost. If the realized linear heat rates — kilowatts per foot (kW/ft) of installed Pu — are only 50 per cent of the 1985 design study target, the added fuel cost is 0.33 mills/kWh. Hence there are economic incentives for achieving target performance, which in this case requires both the 100,000 MWD/T and linear heat rates between 30 and 40 kW_e/ft.

Attainment of this burn-up requires considerable extrapolation of present fuel element performance. Normal burn-up in present light water reactors is close to 20,000 MWD/T, with deviations dependent on the linear heat rates (kW/ft of fuel) differing from roughly 16 kW/ft. This level of performance has been consistent with coolant outlet temperatures to date of about 1,050°F. In the LMFBR, temperatures of 1,250°–1,400°F are considered necessary to achieve linear heat rates of 30–40 kW/ft, and burn-up of 100,000 MWD/T of plutonium.

The development of higher burn-up is achieved by trying out new uranium compounds — by irradiating new rods in a simulated core environment. At the present time, only 12 mixed-oxide fuel rods at 15–16 kW/ft have been irradiated in the experimental facility EBR-II to 50,000 MWD/T. These tests are being extended since it is obvious that much more irradiation will be required before some one promising fuel element can be said to be "developed." Only the experimental breeder reactor EBR-II is available for more irradiation tests, although the operating Detroit prototype, ENRICO FERMI, may become partly available. The first has only 14 to 16 subassembly positions, each of which holds 19 encapsulated fuel rods or 37 unencapsulated rods. Most of the fuel rods irradiated in EBR-II have been encapsulated — have had an added outer jacket of stainless steel for added safety — and thus have experienced lower burn-up than the subassembly would possibly achieve in an operating reactor. In order to enlarge the testing capability and improve the forecast accuracy of the results, new work will have to be done with unencapsulated rods. The decision here shows the typical choice in fast reactor research between "time" and "certainty." Twice as many unencapsulated rods can be irradiated at one time; but they have an increased risk of contamination of the whole set of assemblies in the event of failure of one or a few fuel elements. Since failure of a fuel subassembly has led to the current shutdown of the Fermi fast reactor, this risk is not unimportant.

The compromise research and development program — balancing rapid with safe results — is expected to require approximately 10 years.[2] The program requires several experiments, a number of which are in sequence.

The *first experiment* is to demonstrate the stability of fission over the entire lifetime of the fuel rod. For burn-ups of 100,000 MWD/T, a new, high-temperature, high-capacity test facility — the Fast Flux Test Facility (FFTF) — is to be used which can simulate the fuel rod lifetime exposure in a single year. Many potentially promising fuel rods will be tried.

The most promising rods must be subject to transient power bursts, as a *second experiment*, to verify that they can take sudden reactivity changes without melting or changing shape so badly as to disrupt the flow of coolant through that part of the core.[3] The Atomic Energy Commission has indicated that a large experimental facility, called the Safety Test Facility (STF), is required for this purpose.

A *third round of experiments* has to be conducted on fuel which is subject to extreme conditions. In a closed loop of the FFTF, fuel rod cladding defects are to be created so as to observe resulting changes in coolant contamination and burn-up.

Finally, a *fourth round of experimentation* has to be completed to document the mechanical properties of fuel and clad after partially complete irradiation. The ductility of the fuel undergoes significant degradation after burn-up is partially complete, and it is suspected that the extent of quality reduction is greater under the high-temperature conditions in which breeder reactor fuels operate. In addition, metallurgical changes take place in the fuel as the many fuel element materials

[2] Current plans call for initiating a prototype in 1969 well before this fuel program is complete, so that first cores in near-term prototypes will have conservative fuel rod designs and low burn-ups.

[3] Solid fission fragments can cause from 20 to 30 per cent expansion of the fuel during the irradiation process. Much of this expansion can be contained by making the effective fuel density of compacted plutonium oxide only 75 per cent of the maximum possible density. This allows room for expansion, although it reduces the reactivity of the fuel pin and the thermal conductivity of the fuel. Fuel swelling is reduced at high temperatures (1,000°–1,800°F) in oxide fuels, since nearly all fission gases are released as the target for burn-up is approached.

interact.[4] Observation of these phenomena is necessary for a variety of burn-up conditions, requiring the prolonged and interrupted irradiation of a large number of fuel elements.

The exact time sequence for these experiments has never been specified. Some part of the results will be known by 1974–1975, from use of EBR-II,[5] but these will be limited mainly to burn-up studies of oxide fuels and will advance the development only of oxide fuel elements. This is not the end of fuel element research. Predicted performance of carbide fuels is much closer to the target requirements, so that work on carbides has to intensify after 1975[6] — by continuous use of the experimental FFTF. Other work has to be done after 1975. For one thing, the FFTF closed loops have to be used for the destructive testing of carbide fuel rods after the carbides have been shown to meet preliminary or first-hand irradiation requirements.

The course of the second, third, and fourth rounds of experimentation is controlled by the FFTF, which, according to Atomic Energy Commission predictions, will become available in 1974. After some three to five years of exposure to FFTF test conditions, enough information

[4] The physical properties of fuel and clad undergo changes in the reactor as a result of both irradiation and interaction with the coolant. Sodium bonds, for instance, facilitate heat mass transfer from fuel to clad, but sodium itself forms a ferritic surface layer on austenitic stainless steel clad and removes soldering compounds so that self-welds are formed at various joints in the system. It is necessary to measure the thickness of the ferritic layer to assess the importance of corrosion under various conditions. In addition, changes in creep rupture strength, steady-state creep rates and fatigue strength must be measured for a variety of sodium temperatures and pressures and for different levels of irradiation.

[5] The following table indicates the present status of fuel testing in EBR-II:

Number of Rods Removed from EBR-II

	Prior to 1967	In 1967
Mixed oxides	8	61
Mixed carbides	6	3
Alloy	35	5
Structural materials	500	900

The schedule of future EBR-II irradiations, in the Argonne Laboratory Study ANS 101, shows six subassemblies exposed to 100,000 MWD/T in the next three years, with over 12 additional subassemblies irradiated to levels of 50,000 MWD/T. These plans for this particular facility demonstrate a concentration on near-term developments sufficient for design of oxide core fuel elements for the first large prototype.

[6] Carbide fuel is of interest because of its higher density and thermal conductivity, but there is a major difficulty in this material caused by mass transfer of carbon by the sodium to the clad so as to reduce the structural strength of the clad. Studies have shown that by reducing the fuel density to 75 per cent, up to 30 per cent volume expansion of the fuel is permissible without straining the clad. But these studies have also shown that the penetration rate of impurities into the clad has been temperature dependent, and burn-ups have been limited to the 4.9 per cent achieved in EBR-II.

should be available to develop the first carbide fast fuel element, and the second oxide element. Further development will require the testing of elements in prototypes for several years before final adoption of a rod design. If the first prototype is not operational until 1977, then 1980 would be the earliest time for completion of a second, or improved, carbide fuel element. Either the "optimistic" or "realistic" view suggests that second-round fuel elements will not be available before 1980. Then research begun in 1969–1970 will be completed in some part in 1979–1980, after a decade of work.

The total costs of this program in fuel research can be estimated. The Atomic Energy Commission's forecasts for expenditures for projects in the next five years are used as the basis for this estimate. First, the construction costs of the FFTF are taken to be $87.5 million, the amount of the Fiscal Year 1968 authorization requested of Congress and the Bureau of the Budget by the Commission.[7] Next, the FFTF is expected to cost $18 million each year after it reaches operational status. This amount is scaled up from operating expenses of EBR-II, given that the EBR is a much smaller reactor (62.5 MW_t as compared to 400 MW_t for the FFTF) which operates at lower energy levels and has no closed loops.

Because of the FFTF, the actual costs incurred in the operation of the EBR-II itself are expected to diminish in the 1970's. This smaller test facility will be replaced for the major irradiation tasks, but will continue to be in operating condition as back-up in the event of an FFTF failure. As a standby facility, the average costs of EBR-II will not exceed $15 million per year.

Finally the costs of research on general cladding materials for fuel development are expected to approximate $20 million per year. Included in these costs are those for development of three or four different types of materials, for metallurgical testing outside of a reactor, for thermal flux exposures of new fuel elements, and for safety tests of new fuel assemblies. When the FFTF and several prototypes are in the construction stage, costs are assumed to rise so as to prepare new materials for in-reactor testing. These general reactor testing costs — termed GRT — should increase gradually to $18 million per year by 1969–1970.

The costs of all of these steps together can vary because of unexpected delays in the time schedule or because of unexpected test results. If all tests in the project are on schedule and on target, then costs should be

[7] In testimony in earlier fiscal years, Mr. Milton Shaw of the Atomic Energy Commission had estimated the cost of the FFTF to be $75 ± 15 million. Difficulties in construction and operation of this large fast reactor should result in costs reaching the upper limits of the estimate — as in the case of every other fast reactor, from FERMI in the United States to BR-5 in Russia.

as shown in Table 2.1. This amount of $998 million assumes that the FFTF is built to schedule, is operated to full capacity for four years, and is complemented by full use of EBR-II, if and when operations problems arise. As a matter of course, problems may develop at such a pace and magnitude that the capacity of the EBR will not be sufficient to keep the project on schedule. Additional operating expenses

Table 2.1 Costs for Developing Fuel Elements for the Liquid Metal Fast Breeder Reactor

Category of Expenditures	Fiscal Year (millions of dollars of expenditure per annum) 1966	1967	1968	1969	1970	1971	1972	1973–1980	Total
Fast Flux Test Facility									
Construction	—	—	—	22.0	22.0	22.0	22.5	—	$ 87.5
Design and operation	4.5	7.7	11.1	14.0	18.0	18.0	18.0	18.0	235.3
EBR-II	6.3	7.9	10.4	13.7	15.0	15.0	15.0	15.0	203.3
Materials Development									
Materials research	25.4* (1966–1967)		12.9	14.9	18.0	20.0	20.0	20.0	271.2
In-reactor testing	—	10.0	15.0	18.0	18.0	18.0	18.0	13.0	201.0
									$998.3

* These expenditures are cumulative for fiscal years 1966 and 1967.

associated with delays in finishing the FFTF could then be as much as $40 million. Similarly, poor initial test results could make it necessary to carry out more irradiation studies before finishing the best type of fuel rod — as much as five years longer, which will add $210 million to development costs (assuming annual costs are constant).[8] Total fuel development costs, given the worst possible results at each stage, could then be $1,248 million. As an inner bound, if the FFTF is available in 1973 and only four years are required to develop the first carbide fuel rod, total costs can be as little as $644 million.

[8] If the problems take much more than five years, there is some chance that other reactor types would be sought, since there is a limit to how much the Government would be willing to spend without some indication that the project can meet target estimates of costs.

The Liquid Metal Fast Breeder Reactor: The Development of Core Performance

The performance of the fuel core — the collection of fuel rods making up the critical mass — is not the sum of the measured performance of each fuel rod. The core has behavioral patterns of its own, such as temperature and pressure surges that would not occur in a single isolated rod, and these have to be understood and controlled. Theoretical calculations of energy release from fission must be experimentally verified before demonstration cores can be designed and built.

The focus of the nonbreeder reactor program on thermal or moderated fission behavior left important gaps in knowledge as to the effects of fast or nonmoderated fission on the core environment. Studies are underway to fill the gaps. Capture cross sections — the frequency distributions of absorbed fission particles in the structural materials — are being documented by experiments with the energy range 1 keV to 1.5 MeV; and the resonance and scattering of particles of both fertile and fissile materials in the energy range from 1 keV to 3 MeV are being observed in the same experiments.[9] Continuing at the present rate of progress, the costs of this work are expected to start at $14.4 million per year in 1968, rise to $15 million in 1969 and 1970, and then fall back to between $10.5 million and $10 million thereafter to 1980.

The development of such basic nuclear data is to complement work on the actual use of fuel cores of various shapes and sizes. In the development of nonbreeder reactors, over 50 critical experiments were employed to measure reactivity effects on materials, to record temperature and density changes, and to verify operating procedures with particular core configurations. There has been no such experience to date with plutonium-fueled cores. The primary facilities needed to carry out such tests are the zero power test reactors, ZPR III, VI, IX and the zero power plutonium reactor (ZPPR), with all except this last reactor being reconstructed to take large plutonium cores.[10] They can be made available at a cost. The Atomic Energy Commission has estimated general costs for operating laboratories and test installations for these purposes to be approximately $9 million in 1968 and 1969, and $10 million each succeeding year until 1980.

These studies in reactor physics and core operation are to be accompanied by studies in "safety." Reactor safety analyses are to ensure that major core experiments do not involve "undue risk" to personnel carrying out the work, and are to employ actual studies of hypothetical

[9] The locus of such research and the present state of knowledge are discussed in Note 1 of Appendix B.

[10] Cf. Appendix B, Note 2, for a description of these reactors.

extremes in the operation of fast reactors. Most of this latter research in the liquid metal reactor program is in simulation of accidents in the core to show the environmental effects of these accidents. Two experimental installations, SPERT and PBF, are currently the center of tests of this type on nonbreeder reactors and can be used for breeder experiments as prototypes approach the construction stage. The costs of operating them for this purpose will be approximately $8 million in 1968, $9 million in 1969, $15 million from 1970 to 1972, and $25 million each year thereafter to 1980, at a cumulative expense of $262 million. These will not provide enough experiments in extreme operating conditions; the Safety Test Facility will be constructed at an additional cost of $66.5 million to provide the needed capacity.[11]

Plans are practically complete for the necessary experiments with core performance to meet design specifications for a 1980 sodium fast reactor. The total forecast costs are $623.6 million before 1980. But what if good or bad results call for significant departures from these plans? Initial research might show that the fuel element can be put together without extensive refinement of existing materials, so that substantial reductions in program size and expenditures are possible. The project plans might also be altered because of poor results at the early stages — demonstrating the need to fabricate exotic new cladding materials — so that costs have to exceed, and performance has to lag, expectations. The desired development can be achieved at some level of costs even after repeated failure, but the question is how extensive a departure would be required from expected research costs.

The engineering analyses and reviews lead to the conclusion that the program either will be followed closely or else departed from significantly.[12] The first round of observations of prolonged irradiation might show that oxide rods are capable of attaining the desired burn-up with sodium outlet temperature at 1,200°F, so that there is no necessity

[11] The development of an economic fuel cycle does not stop with achievement of a reliable and efficient fuel core. The fuel elements have to be fabricated in volume, and reprocessed after coming out of the core. According to General Electric analysis, five alternative approaches to fabrication and reprocessing have been identified for oxide fuels alone (GEAP-4418, pages 4–18). These have to be tried by actually processing fuel and then observing its performance within an experimental or demonstration core. The most promising approaches would be carried through to preparation of working cores. The former preliminary steps would probably cost $32 million from 1964 to 1980, and the latter would cost $30 million in the construction of small-scale alternative manufacturing plants similar in capacity to EBR-II.

[12] Cf. three analytical reviews of fuel development problems: (1) M. C. McNally *et al.*, *Liquid Metal Fast Breeder Reactor Design Study*, General Electric Rept. GEAP-4418, Part 4 (January, 1964); (2) Argonne National Laboratory, *LMFBR: Summary of Preliminary Plans* (March, 1967); (3) American Nuclear Society, *National Topical Meeting on Fast Reactors* (April 10–12, 1967).

for an extensive uranium carbide program. In contrast, five years of irradiation experience could lead to the complete abandonment of a uranium oxide fuel core because of exceptionally poor longevity or buckling of cladding materials. There would have to be more safety work, more capacity in STF, and five more years of testing new cores and combinations. Each extreme seems as likely as the design program. But the first would cost only $241 million, and the second would cost $883 million before completion of successful tests of a carbide-fueled core in 1985.[13]

The Liquid Metal Fast Breeder Reactor: Components and Systems Development

An exception to the statement that fuel development is the major LMFBR problem could well be raised by those having had experience with liquid metal-cooled systems. Even when fuel cores produce energy reliably, liquid sodium heat-transfer networks are not operable in a large number of cases. There have been problems with coolant blockage and radiation, and with instrumentation for detecting these problems — all related specifically to the properties of sodium.

Sodium absorbs heat so well that the reactor outlet temperature can easily reach target temperatures of 1,200° to 1,400°F without pressurization; but components and piping have to operate at red heat at these temperatures and have to undergo extreme thermal expansion from temperature transients of up to 800°F. Sodium is highly interactive with air or water at these high temperatures and absorbs so much radioactivity that piping materials corrode. An accident opening the primary heat transfer tubing could be followed by destruction of most materials in the immediate environment.[14]

The most serious of the materials problems is the development of piping that can withstand the extremes of thermal stress and corrosion. So far, stainless steels have been used successfully when temperatures have been limited to 1,000°F but not at the design target of 1,400°F, so new alloys may have to be produced to gain the additional 400°. The first experiments in this direction promise to upgrade the known steels,

[13] The estimates are constructed by first deleting carbide development from the program entirely, and second increasing both its rate and scale so as to make uranium carbide rods the only fuel source for fast reactors in 1980. The numbers are subtractions from or additions to the program expenditures, in keeping with the planned time schedule terminating in 1980. But in the case of "high" costs, five additional years of core operation experiments will cost $200 million; additional safety tests, $40 million; and additional operation of STF, $20 million.

[14] Cf. American Nuclear Society, *National Topical Meeting on Fast Reactors*, pp. 6–37 to 6–56.

and the second and following experiments to search for an appropriate material in the more exotic metals if the first round does not succeed.

After some progress has been made on materials, there is the problem of components for the reactor and for system control. Pumps, valves, seals, bearings, piping, heat exchanger piping, and purification equipment are all exposed to high-temperature sodium. Most of these items have been built to 50 per cent of scale and operated in test facilities at 90 per cent of design temperature; new, larger units have to be produced to test for full-scale reliability at required performance standards. The problem again has an either-or solution: build larger units of the same design as those now used, or start over with an uncertain design that has higher theoretical performance. For example, the mechanical seals used in presently operating sodium reactors have a lubricated face, and can serve at 1980 design levels; but "the use of a lubricant requires an elaborate design to prevent the lubricant from contaminating the sodium . . ."[15] while a dry-face seal has a much less elaborate design and the promise of more reliable long-term performance. The first is well developed, requiring enlargement and improvement in operating life, while the second has never been adapted to a sodium environment, so that it has a smaller chance of achieving the required higher standards.

Decisions will be made on which path to follow during experiments at the components test facilities, operated by the Atomics International Company in California. First-of-a-kind pumps, valves, piping arrangements, etc., will be installed in a sodium loop fired by a gas burner.[16] The approach to all components is to build one and test it under conditions as similar as possible to those in the 1980 reactor, and then discard it for another model if it fails to work under those conditions. The costs of the Atomics International facilities, and of other component development projects that are part of the experimental approach, can be expected to be very high.

The first category of expenditures covers the operation of the Sodium Components Test Installation (SCTI). This plant is to test new models of particular items of equipment in time for the construction of 300 MW_e prototypes (the 300 MW_e plants have to have components appropriate for the 1,000 MW_e plant to serve as a 1980 demonstration project). Since there are numerous contending designs for the sodium subsystem, each of the first few prototype reactors should have different heat exchangers, pumps, etc., so that the test facilities have to be large enough to carry out tests on a number of items at the same time.

There are a number of research problems to be investigated in finding all of the best components. The technology of sodium has to be

[15] *Ibid.*, pp. 8–26.
[16] *Ibid.*, pp. 8–35, *et seq.*

investigated in more detail throughout this period. The sodium-water reaction has never been thoroughly explained or understood; in addition, problems from mass transfer and other impurities in sodium remain to be solved before commitments are made to use particular metals. A heat exchanger must be made that will work without materials deterioration during the plant lifetime; this is a project with priority over the testing of other components because the test loops themselves have to have such exchangers. These two types of projects are related: problems in the exchanger such as extinguishing sodium fires and containing sodium-water reactions and rapid thermal transients can be dealt with when more is known about the behavior of sodium.

The components themselves must be beyond the design stage and standards for performance must be established before testing can begin. This phase of the program should be completed in five to seven years, with the tests verifying the usefulness of the available hardware toward the end of that period.

The critical component development problems — compared to component research problems — are in finding safe and low-cost containment for the reactor. The vessel and shielding of fast reactors built to date have made up a "complex structure which generally has been more difficult to analyze than the containment structures used for the [nonbreeder] reactors"[17] because of the special need to prevent the escape of sodium. The way to a low-cost and safe structure seems to be to divide it into two parts, one enclosing the core and primary heat exchanger, the second enclosing the entire plant, with both parts in reinforced concrete. This would seem to function, but there is still work to be done on controlling cracks and leaks, particularly when internal heat exceeds design conditions. The results here determine the economic advantage of large fast reactors:

> The significance of such cannot be over-emphasized . . . this can mean the difference between a row of $\frac{7}{8}''$ diameter bars spaced at 10″ to 12″ and a double row of $1\frac{1}{2}''$ bars spaced at 4″ to 5″ on center.[18]

The research program to provide answers here begins modestly but will have to expand.

The total costs of components development depend on the extent of first-round answers on solving these problems. Moderate success would consist of achieving target performance with the third or fourth test components — with new metals or radical designs in a quarter of the cases. The costs of this level of achievement generally are expected to add up to the annual expenditures shown in Table 2.2.

[17] *Ibid.*, pp. 6–46.
[18] *Ibid.*, pp. 6–51.

Table 2.2 Components Development Expenditures

Fiscal Year	Millions of Dollars
1966	6.2
1967	6.7
1968	12.3
1969	18.1
1970	16.0
1971	16.0
1972	16.0
1973	16.0
1974	16.0
1975	16.0
1976	16.0
1977	16.0
1978	16.0
	172.4

After the many components have shown reasonable technical performance in a test facility, they must be put together to make up systems if they are to be convincing in their demonstrations of economic operations. There are currently two sodium circulation designs and at least three different heat exchanger designs so that — with four core configurations — there are 24 possible prototype designs. All of them of course will not be constructed — the test facilities should produce results that eliminate most of them prior to any construction. But this is not the case for the four core types and two sodium circulation designs. A number of alternative prototypes will be needed to choose the final LMFBR design, because there is no way to evaluate a core's behavior short of building a model of that shape and size.

One of the major problems with liquid metal designs is that there are significant differences in the fission behavior of small fuel cores and the larger cores for the 1980 full-scale reactor. Therefore, small experimental cores are not adequate for proving the worth of configurations in large reactors, and there is the need to build a number of large prototypes. The costs of these prototypes will exceed conventional nuclear costs for comparable-sized plants. The upper limit on the difference must be 100 per cent because otherwise there would be no cooperative development programs in which costs were shared equally by the government and private utilities. For a 300 MW_e prototype, this sets a limit on research costs of about $300 per kilowatt and for a 600 MW_e plant the upper limit on costs would be $225 per kilowatt.

These research costs are expenditures beyond the costs of present water reactors, and come to approximately \$30 million for a 300 MW_e plant and \$45 million for a 600 MW_e plant.

Total requirements for "proving out" fast reactors could well include three prototypes and one large demonstration reactor, given the number of promising core configurations and methods of circulating sodium. The schedule for best evaluation of alternative configurations requires construction of prototypes in 1970 and 1972 at the 300 MW_e level, a 600 MW_e plant in 1976, and the demonstration plant itself soon after some experimental results have been produced.[19] Each prototype would take six years from contract commitment to operation, if the construction of LMFBR components is given higher priority with reactor manufacturers than present conventional reactor components.

This is a tight schedule. Commitments for the 600 MW_e plant have to be made before there is very much operating experience with the first prototype. But this is the quickest and cheapest path to moderate success with operating systems. Total costs would be \$379 million, as a result of three classes of expenditures; first, approximately \$139 million would be used for systems engineering and operations, at the rate of \$10 million per annum; second, \$65 million would be spent on prototype research designs; third, \$175 million would be spent on constructing these plants from 1970 to the early 1980's.

The whole of the program to produce a sodium fast reactor can possibly produce more and varied research results than can be known and evaluated at the present time. Knowledge will be accumulated from other research projects in the energy industries, and some of it may well offer cheap new means of cladding fuel. But other new findings elsewhere may make it clear that presently promising research approaches will ultimately not work, so that some of the present alternatives will be foreclosed or rendered obsolete. What is known of the possible range of results, however, is reflected in the range of costs shown in Table 2.3. With exceeding success in initial operation of each prototype after good first results on each experiment, costs could be as little as \$1.4 billion for the whole project. With middling success — not having to start anew with new materials after testing early models of

[19] The Atomic Energy Commission has tentatively indicated interest in three prototypes spaced several years apart. It is conceivable, however, that four might appear as a consequence of utility pressure to support a more competitive environment. If General Electric were to obtain support for the first prototype, for instance, it is likely that Westinghouse would press a second group of utilities to obtain Atomic Energy Commission funds for a second prototype very shortly thereafter. Combustion Engineering, Atomics International, and Babcock and Wilcox would press for other prototypes on the quite valid grounds that to be left out at this stage is to be left out entirely.

Table 2.3 Development Goals and Costs for the Liquid Metal Fast Breeder Reactor

Objectives	Present Position	Program to Termination of Project	Equipment Required but Not Yet in Place	Costs (millions of dollars) Low Cost	Design Cost	High Cost	Time
1. *Fuel Rod Performance*							
a. 100,000 MWD/T at 800°–1,200°F for coolant, and producing 5 to 25 kW/ft	a. Irradiation of fuel elements in EBR-II, SEFOR to 100,000 MWD/T is underway. Initial results show "promise" beyond 50,000 MWD/T, but experienced "high" failure rate. Present facilities extremely limited.	a. Continuation of irradiation in existing fast facilities and FFTF; three-year exposure of some fuel elements required.	Fast Flux Test Facility (FFTF)	644	998	1,248	1980
Fuel Clad Performance							
b. Develop SS to take radiation damage and transient over-power to level required for a.	b. Initial irradiation of clad in fast flux to 50,000 MWD/T concluded. Treat tests of power transient effects underway, with initial tests concluded successfully.	b. Documentation of resistance of 316SS to mass transfer in dynamic sodium. This is a six-year project (two for construction, four or more for testing).	Safety Test Facility (STF)				1975
Other materials to show same level of promise as stainless steel now shows.	Irradiation in thermal flux beyond 125,000 MWD/T achieved successfully.						
2. *Fuel Processing*							
Development of a single process, of the five proposed, that produces mixed-oxide fuel to performance. Experimental research to establish cheapest recovery methods.	The Purex process on a single cycle works for low radiation.	Small-scale to pilot-scale experimentation with all processes.		62	62	62	

3. *Subsystem: Core Performance*							
a. Determination of critical mass. b. Reactor geometry for maximum neutron economy and stable operation.	a. No Pu-fueled fast assemblies in operation; no material holding a fast assembly together has been developed.	a. Testing of assemblies for criticality; mockup of 1,000 MWt core with control configurations.	ZPPR-6/28-Mockups and clean critical-expts for a, b, c. ZPR-6, ZPR-9, Reactivity Coefficients for c. Fast Spectrum measurements for c.	241	624	884	1978
c. Energy spectrum under wide variety of conditions.	c. Reactivity coefficients differ by factor of two or three. c. Irradiation program underway in 1963 adequate, but time is needed.	c. Measurement of reactivity and Doppler in one isotope at a time in various neutron spectra, all in large critical facility with heated core. Doppler measurements in SEFOR and FERMI.	ORELA-1/69, Fission and Cross Section measurements for a, b, c, d.				
d. Impact of bubbling and fouling in the core.		d. Data from TREAT tests and FFTF on sodium voiding.	ALN NEUTRON GENERATOR -1/69 for cross section measurements, a–d.				
e. Meltdown characteristics.	e. Mark II core of EBR-I yielded meltdown information.						
4. *Subsystem: Components and the Behavior of Sodium Coolant*							
a. Components such as pumps, heat exchangers, valves, seals and bearings, purification equipment, liquid metal instrumentation, contaminant vessels.	a. Materials are adequate for near-term and most long-term objectives.	a. Sodium components development program carries these to the prototype. Test loops to evaluate corrosion and heat transfer (SCTI).	STF (6/73)	142	172	292	1978
b. Data on sodium-to-steam heat transfer characteristics of the system.	b. Operating experience with sodium up to 1,000°F. Uncertain as to operating characteristics of primary loop between 1,000°F and 1,250°F.	b. Measurements of creep rates, fatigue strength at 1,400°F. b. Experiments to test proclivity of system to self-weld above 1,000°F.					
c. Data on basic sodium properties; data on basic sodium–water–air interaction.	c. Experiments with boiling at AI and APDA.	c. Develop sodium purification methods and knowledge of material compatibility with total system.					

Table 2.3 *Continued* Development Goals and Costs for the Liquid Metal Fast Breeder Reactor

Objectives	Present Position	Program to Termination of Project	Equipment Required but Not Yet in Place	Costs (millions of dollars) Low Cost	Design Cost	High Cost	Time
5. *System Development*							
Instrumentation and control of the entire system.	FERMI shut down. EBR-II in operation. SEFOR in test and initial operation.	Thermal cycle analysis to establish temperature rises throughout the system, and to set optimal core thermal power. Optimization of Fuel Cycle.	FFTF STF	279	379	741	1985
Operation of prototypes in accordance with design specifications.		Computer simulation of system, to include reactivity coefficients; simulation of accidents. Integration of components to achieve system capability. Construction of prototypes to test subsystem integration, and to estimate costs of operation.	4 Prototypes 1 Demonstration Plant				
				$1,378	$2,235	$3,227	

the first design — costs of research are expected to be close to $2.2 billion before the 1,000 MW_e demonstration reactor is in operation in the early 1980's. But with equally probable poor results at each step, including those requiring reversals of early decisions on the best fuel rods, complete government payment of all costs on first and second prototypes, and the necessity for another 300 MW prototype at $90 million, the total project expenditures would exceed $3.2 billion by the late 1980's.

The Gas-Cooled Fast Breeder Reactor: Fuel Element Development

A fast breeder using a gas as the coolant has claim, at the design stage at least, to consideration as the subject for a large-scale development project. Helium has been used as the coolant in nonbreeder reactor experiments and in the full-scale prototype nonbreeder reactor at Peach Bottom, Pennsylvania, so that some experience has been accumulated with components in a gas environment. The heat exchangers, circulators, valves, etc., needed in a gas fast reactor are identical to those required for the advanced nonbreeder prototype to be built at Fort St. Vrain. However, the gas breeder cannot be constructed at present because fuel elements, core performance, and systems operation have not been developed to breeder standards. There are additional problems caused by a lack of operating experience with gas reactors in the United States, so marked that the gas breeder would probably require more time than the liquid metal breeder from the beginning of fuel rod research to the completion of a 1,000 MW_e demonstration plant.

The fuel element requires a design for the fuel pin and a clad material compatible with both plutonium fuel and helium coolant. It is conceivable that knowledge of fuel pin behavior obtained in the liquid metal fast reactor program will be useful for pin design for a gas fast reactor. If this is the case, then development of fuel for the latter system could concentrate on the testing of clad in the helium environment, by installing gas loops in the FFTF and by constructing a Gas-Cooled Fast Reactor Experiment plant (GCFRE) to operate a full-sized fuel core in a helium circulation system.[20]

[20] The GCFRE would take three years to produce burn-ups of 100,000 MWD/T, compared to four years in EBR-II but one year in FFTF. It would have 200–500 liters of capacity. It could very well be used to develop a reliable fuel, but not to explore a large number of alternative fuels in a short time interval. The limited search seems reasonable, given the present state of gas technology, but could result in a long delay in the development of the lowest cost fuel element (except by a chance choice of the best material first).

The gas fast breeder might be the only concept on which research takes place. The fuel pin work then would proceed as in the sodium fast reactor project, given that this part of the project produces results applicable to either fast reactor. Corresponding work on cladding materials would be held back at first by the lack of a fast flux *helium* environment in which to test a complete fuel element; rapid construction of FFTF (for closed loop testing of elements) and GCFRE (for both fuel element and subassembly testing in the appropriate gas environment) would solve this problem at costs of $87.5 million and $30 million, respectively. The FFTF would become operational in 1974, as in the liquid metal program, and the GCFRE should be available at the same time, given four years for construction after submittal of the preliminary design and that the design is complete in 1970. Operating costs would be $235.3 million for the FFTF over the 1966–1980 period (as noted for the LMFBR) and another $197 million for the GCFRE for the same period (equal to those for EBR in the liquid metal program).[21] The last of the three major facilities useful for backup to research is Experimental Breeder Reactor EBR-II. When the other two are available, there is little need for EBR-II; as a result, operating expenses for this reactor would be negligible after 1969.

Even with all efforts focused on the helium gas environment, the period for fuel development would be at least two years longer than for sodium breeder fuel. The gas project would have a smaller number of test positions which implies slower accumulation of test results. As a consequence of this longer research period and the need for the highly specific GCFRE, the expenses for fuel development shown in Table 2.4 are $117 million greater than for the same parts of the sodium program.

An optimistic estimate of fuel costs can be constructed after assuming that the element is developed on a sodium-type time schedule — or four years earlier than appears likely — and that the FFTF costs are close to the optimistic estimates in the liquid metal reactor program. These assumptions are going to be realized if each part of the many construction projects is completed on schedule and all experiments come out with "best forecast" results. The costs would be the low LMFBR costs of $644 million (as in Table 2.3) plus $30 million for GCFRE construction, or $674 million.

A pessimistic estimate is obtained from assuming higher-than-expected construction costs for the FFTF and GCFRE and an even longer period of development. The FFTF could cost as much as implied by poor construction and operating results in the LMFBR program (as shown by high costs in Table 2.3). The same results on the GCFRE

[21] This is comparable to Atomic Energy Commission spending on one major fuel program, such as uranium carbide, at the present time.

Table 2.4 The Costs of Fuel Development for the Gas-Cooled Fast Reactor

A. FFTF construction costs are $87.5 million.
FFTF design and operation expenses:

1966	1967	1968	1969	1970	1971	1972	1973–1980	Total
4.5	7.7	11.1	14.0	18.0	18.0	18.0	18.0	$235.3 million

B. Fuel Materials Development (includes general reactor technology costs for fuel):

1967	1968	1969	1970	1971	1972	1973–1982	Total
35.4	27.3	32.9	36.0	38.0	38.0	38.0	$587.6 million

C. The Gas-Cooled Fast Reactor Experiment:
Construction costs of $30 million.
Design and operation expenses:

1969	1970	1971	1972	1973–1980	1981–1984	Total
2.0	15.0	15.0	15.0	15.0	7.5	$197 million

D. Experimental Breeder Reactor-II
Operation expenses:

1966	1967	1968	1969	1970	1971	1972	1973–1980	Total
6.3	7.9	10.4	11.7	—	—	—	—	$36.3

E. The total of A, B, C, and D is $1,056.7 million.

would extend costs of that facility another $30 million. Then total fuel element development costs would be $1,338 million — $1,248 million for the orthodox FFTF and EBR programs, with an added $90 million for the special gas environment in the GCFRE plant.

Rather than an exclusive gas program, there could be projects to develop both liquid metal fast breeders and gas fast breeders. The two projects could then be managed to use certain facilities at the same time. An entirely separate FFTF would not be needed for gas research, but rather only gas loops added to the plant at costs of $15 million for construction and $3 million annually for operation. (Since there would be fewer test positions available,[22] it would take roughly five years longer to finish the same gas fuel element tests, however; the additional

[22] Half the capacity in the FFTF is to be for fuel pin development and testing rather than for complete element and fuel subassembly testing. If the total number of gas test positions is reduced from six to four to accommodate two programs, then 50 per cent more time or an additional five years of irradiation is required to develop the fuel.

years of FFTF operation at half capacity would involve expenses of $9 million per year.) GCFRE operating expenses would be $7.5 million per year for each of these additional years as well, and fuel materials experiments would again cost $20 million per year or $100 million more. The total additional costs for gas as incremental to liquid metal would then be $263 million. Base costs for gas would still have to be incurred for the GCFRE, testing gas exclusively, and would still come to $207 million. Hence the total expected costs for a gas reactor as a second project would be $470 million. If "research breakthroughs" made it possible to complete this stage on time — five years earlier — costs would be $300 million. If, on the other hand, the usual delays were compounded by adding gas to liquid metal research in a single set of laboratories, five to seven years and $227.5 million would be added to the costs which would result in total incremental expenses of $698 million.

The Gas-Cooled Fast Breeder Reactor: The Development of Core Performance

The fuel cores of gas-cooled commercial fast reactors cannot be constructed until three research projects show at least preliminary results — the same projects taken on for the same reasons as in sodium-cooled fast reactor development. First, data have to be collected establishing the frequency distribution of neutron energy in a fast system with helium as the coolant so as to plan the size and shape of the critical mass. Second, initial experiments on critical mass performance have to be conducted on mock-ups of the mass. The third step is to build full-sized cores of different designs in experimental reactors, to test performance under both normal and induced disaster conditions. Then enough results should be accumulated to choose promising core configurations, to predict the behavior of such chosen configurations, and perhaps to build one or two of them.

Most parts of this plan are the same as those in the development programs of other fast reactors. Data collection and the subsequent construction of mock-ups are done in the same manner and at the same cost as in the sodium breeder program, even though the coolant is not the same. The project to test critical cores may differ in content — corrosion tests of the assembly in a sodium environment may not be duplicated for the gas-cooled cores because corrosion is not a problem of the same magnitude — but the capital and personnel required to carry out the testing program are of the same order of magnitude. Costs for fuel core development in a gas fast reactor are estimated to be the same as in the steam-cooled reactor, except that the chemistry and physics of helium are relatively well known as compared to sodium, so that $100 million of materials expense can be foregone on the gas programs that are part of the sodium program. Low costs are $141 million, "target" or expected costs $424 million, and high costs are $784 million.

In the case of a gas-cooled program incremental to a liquid metal fast reactor development program, the costs range from $148 million to $252 million with expected results at $178 million. These estimates depend on a number of assumptions. Gas research could be somewhat delayed and put behind the work done on the sodium core; then gas expenditures would be limited to the added costs of operating the zero power and safety test equipment for the extra gas experiments. It is reasonable to assume that several core spectra measurements per year and at least one mock-up are necessary so that the costs of the first and second steps of an incremental gas program are about $1.8 million per year. In addition, comparable reduced expenditures could be expected for the third step — that part of the program devoted to analysis of core operations. Even though a part of research should be of general use to either coolant system, there is no substitute for reactor facilities devoted exclusively to the gas environment; a total of $1 million per year seems reasonable for work in such a ZPPR facility. Finally, costs for safety analysis in an incremental operation are closer to those for the single gas project, because here the programs diverge (except for the development of fast control mechanisms, projects to observe fission behavior at limit temperatures are different because the coolant significantly affects such behavior). These last costs are assumed to be twice as great for two projects as for one.

The Gas-Cooled Fast Breeder Reactor: Components and Systems Development

The extensive program for developing components for high-temperature gas nonbreeder reactors is about to bear fruit in the installation of pumps, heat exchangers, and gas piping in the Fort St. Vrain reactor. Given good results, components development expenditures will be at the modest levels expected by industrial sources: the Gas Heat Exchanger for $1–2 million; Gas Circulator for $1–2 million; PCRV research for $3–5 million; the Full Scale Mock-up for $1 million; with the total coming to $6–10 million. If $5 million are required to upgrade all other components to a level of quality slightly higher than for the advanced nonbreeder, maximum expected costs are only $15 million. In addition, control rod drives, valves, tubing, etc., will cost in the neighborhood of $5 million to develop. Hence total components development should lead to expenditures of about $20 million. In the event of extreme difficulty, costs could be twice as much as this projected amount. Given the rather straightforward development problems that lie ahead, this would be difficult to achieve. Similarly, if all goes extremely well, costs might be as little as $6 million for major components and $4 million for minor components, or $10 million in total.

Even though gas coolant technology is in the process of being demonstrated in the United States in high efficiency nonbreeder reactors, no gas breeder system is now in operation or even at the prototype stage. As a result, systems development will be longer and more expensive than for the other type of breeder reactor. The gas prototype research compares with that in nonbreeder reactors in the late 1950's; there will have to be a number of prototypes so as to "learn by doing" actual plant construction. The boiling water reactor program had six reactors under 100 MW_e, and only one in the 300 MW_e range, en route to competitive status. Nonbreeder gas reactors contemplate a scaling up from Peach Bottom at 35 MW_e to 300 MW_e at Fort St. Vrain; this size increase will yield much information on whether further increases are feasible in the form of large gas-cooled fast reactors. But at least one prototype plant in the 50–100 MW_e range, specific to the fast gas system, is required to show technical reliability, and approximately three plants in the 300 MW_e range are needed to test the second round of improved components. Then a 600 MW_e size plant would evaluate another round, this time evaluating the full-scale components and core assembly for a 1,000 MW_e demonstration plant.[23] The costs of these plants and related research are shown in Table 2.5, based on the research portion of the first plant costing $300 per kilowatt, of the three prototypes costing $200 per kilowatt, of the full-scale prototype costing $160 per kilowatt, and the demonstration plant costing $65 per kilowatt.

These are costs for a program that does not achieve a startling breakthrough at any stage. They can be much higher if presently unexpected technical difficulties are encountered in the program — such difficulties as have been endemic in Peach Bottom (which had an initial cost estimate of $700 per kilowatt and was finally constructed for approximately $1,380 per kilowatt). Certainly, actual costs of twice the budget originally estimated are not unknown, at least on the earliest prototypes;[24] but given that the advanced gas converters are now providing

[23] This is to reproduce and expand on the first steps in the gas nonbreeder program: one prototype similar to Peach Bottom will be essential to learn the technology of the system, and three prototypes on the scale of Fort St. Vrain are required to proof-test all the gas components; the full-scale prototype will reproduce the last step in the gas advanced-converter program (a step not yet begun).

[24] Cf. Appendix A for a history of such estimates.

Table 2.5 Costs for Gas Fast Breeder Prototype and Demonstration Reactors

Year	Prototype Design	Prototype Construction	General Systems Development
1971	10	15	29.5 (1968–1971)
1972			4.8
1973	15	60	5.0
1974			5.0
1975	10	60	5.0
1976			5.0
1977	5	60	5.0
1978			5.0
1979			5.0
1980	10	96	5.0
1981			
1982			
1983			
1984	10	65	
	60	356	74.3

the equivalent of "late information," the high range of probable costs is twice that shown in Table 2.4 for the 50 MW_e and first 300 MW_e plant but only 25 per cent greater than that shown for all subsequent plants. These costs come to \$145 million more for plant construction than shown in Table 2.4; in addition, there is likely to be more research design and experimentation expense, primarily in the early plants, adding \$40.3 million to prototype costs. Total plant costs at the high end of the range could be as great as \$675 million.

At the opposite end of the range of results, it is possible that costs may be significantly less than expected. Those immediately concerned with development of this reactor state that it is possible to go directly from the 50 MW_e experimental plant to a 1,000 MW_e prototype plant, given that gas components exist from the Peach Bottom–Fort St. Vrain program and that core performance is independent of the size of prototype (whereas the LMFBR must face *increased* positive reactivity and temperature from sodium voiding of the core in an accident with *increasing* size). Then good research results early in the program would make it possible to build only a 50 MW_e and a 1,000 MW_e plant, which would make costs of all plants equal to \$80 million. There would be accompanying low-level costs of design research and experimentation of \$74 million, so that plant costs would total \$154 million under the best circumstances and results.

Table 2.6 Development Goals and Costs for the Gas-Cooled Fast Reactor Program

Objectives	Present Position	Program to Termination of Project	Equipment Required But Not Yet in Place	Research Costs (millions of dollars) Low Costs	Design Costs	High Costs	Time
1. *Fuel Rod Performance* a. 100,000 MWD/T at 800°–1,200°F for coolant, producing 5–25 kW/ft.	a. No operating experience of significance — Peach Bottom just begun.	a. Irradiation in fast reactors — FFTF if possible and GCFRE if built, possibly Dounreay.	FFTF GCFRE	674 (300)*	1,055 (470)	1,338 (698)	1984
Fuel Clad Performance b. Develop clad capable of high pressures (1,000 psi) and temperatures (1,300°F).	b. No relevant experience.	b. Irradiation testing of fuel rods with different cladding materials and surface roughening in gas environment in FFTF and GCFRE.					
2. *Fuel Processing* Develop an economically attractive recovery method.	Aqueous, volatility and pyroprocessing techniques being developed by AEC for other fast fuel elements.	Small scale to pilot plant experimentation with all processes.		30 (30)	30 (30)	30 (30)	
3. *Subsystem: Core Performance* a. Critical mass.	a. No Pu-fueled fast assemblies currently in operation in U.S.	a. Testing of critical assemblies, test mockup of 1,000 MW_e core with control configuration.	a. ZPPR-type assembly.	141 (148)	424 (178)	784 (252)	1982
b. Reactor geometry for maximum neutron economy and stable operation.	b. Uncertainty on magnitude of coolant coefficients.	b. Critical experiments to determine coolant coefficients.	b. ZPPR-type experiments.				

c. Energy spectrum under wide variety of conditions.	c. Reactivity coefficients poorly known in fast gas systems.	c. Measure reactivity coefficients in GCFRE, critical experiments; determine cross sections.	c. Cross section measuring devices (ORNL, ANL machines).				
d. Impact of loss of coolant on safety.	d. No information.	d. Tests in STF, PBF type facilities.	d. Experiments in GCFRE.				
e. Meltdown characteristics.	e. No information.	e. Theoretical investigation and FARET-STF type experiments.					
4. *Subsystem: Coolant System*							
a. Components such as gas circulators, heat exchangers, valves, alternate cooling systems, prestressed concrete reactor vessels, etc.	a. Small-scale components in use at Peach Bottom. 300 MW_e size components in design for PSCC reactor (thermal). PCRV prototype tested.	a. Components development program.		10 (10)	20 (20)	40 (40)	1979
5. *System Development*							
Instrumentation and control of entire system. Operation of a prototype in accordance with design specifications.	Peach Bottom just beginning full-power operation. No fast gas reactors in existence.	Build 3 prototypes in 300 MW_e range, one at 600 MW_e, and then a 1,000 MW_e demonstration plant.	3–Prototypes (300 MW_e) 1–Prototype (600 MW_e) 1–Demonstration Plant (100 MW_e)	154 (154)	490 (490)	675 (675)	1987
				1,009 (642)	2,019 (1,198)	2,867 (1,695)	

* ( ) assumes that the gas-cooled program is incremental to a full-scale liquid metal fast breeder program.

The costs for all aspects of this gas development program are expected to come close to $2.0 billion, if all specific projects are completed on time with reasonable success (as shown in Table 2.6). With splendid success, the time schedule could be shortened to have a demonstration 1,000 MW_e gas fast reactor in operation in 1983 (as planned for the liquid metal reactor) after having spent only $1.0 billion. But there are good reasons for the final results in 1987 costing as much as $2.8 billion.

Perhaps some part of these costs could be spared by attaching a gas breeder program to the trailing edge of a sodium program. The costs of "adding on" are shown in parentheses in Table 2.6; they indicate that this breeder might be developed for little more than $640 million, or — after research results have been disastrously poor — as much as $1.6 billion. These gains from consolidation are achieved by cutting into the number of groups working on more or less the same problem, and making the results of one experiment serve both breeder programs. These possibilities are matters of strategy.

Strategies on the Costs of Research

The research programs are each geared to development of a reactor type determined by the coolant — liquid sodium (LMFBR) or gas (GCFR). Each program is divided into major problem areas to be dealt with before the construction of any plant is begun. The problems are in fuel element operation, core performance, and equipment performance. The approach to solving them can be to choose one reactor type and then to spend money until there is assurance that a reactor demonstrating that concept can be made to operate consistently. A quite different approach is to conceive of the development of fast breeders as problem-solving in fuel element performance regardless of the coolant, followed by specific development of equipment subsystems of particular types of reactors. For instance, the development of fuel materials — be it oxide or carbide alloy — is a problem common to all reactor types, but the components in the reactors have quite specific development problems. The second approach would possibly imply one development program starting with fuel research and continuing with separate core performance and components research projects in each coolant environment.

An assessment of the first approach has to begin by comparing the costs of research between the liquid metal and gas research projects. There are any number of possible "costs of research," however; some assumptions are necessary to make the comparison possible. Assume that the subproducts of research in the two projects are roughly the same — so that any cost difference is not a result of planned differences in reactor performance. A positive view of the project requires that the chances of "high," "target," and "low" costs at any stage be assumed to be roughly the same. Even though the design studies may support and stress "target" costs, "high" costs must be as likely; it is optimistic to assume that "low" costs are as likely as the other two. Assume also that each of the four stages of any project can have "high," "target," and "low" results independent of those at other stages. The achievement of "target" costs in core performance research does not preclude "high" or "low" costs on components development. Then there are 81 possible combinations of "high," "target," and "low" costs for the four stages in each development project, and these make up a frequency distribution of costs of research for that project.[25]

Table 2.7 Estimated Costs of Two Fast Reactor Development Projects

	Costs of Research (billions of dollars)	
Probability of Costs Less Than Those Shown	Liquid Metal Fast Breeder Reactor	Gas-Cooled Fast Breeder Reactor
0.01	1.3	1.0
0.10	1.7	1.3
0.20	1.8	1.6
0.35	2.0	1.7
0.50	2.2	2.0
0.65	2.4	2.1
0.80	2.5	2.3
0.90	2.8	2.5
0.99	3.2	2.8

The characteristics of these frequency distributions for the liquid metal and gas-cooled reactor projects are shown in Table 2.7; comparison is straightforward. The estimated costs of the liquid metal program are greater for any probable level of occurrence than for the gas program. The liquid metal program is supposed to be complete in 1985, while the gas program incurs some expenditures stretched out almost to 1990, so that the present discounted values of the cost differences are much greater than the undiscounted differences shown in the table.

[25] The combination of these cost results HTL on four parts of the project — (a) fuel rod, (b) core performance, (c) components, (d) systems development — include H(a), H(b), T(c), T(d), HHHT, LTHT, and so on. There are $(3)^4$ such combinations if independence is assumed.

The research effort can be organized according to the second approach, so that there is a "mainstream effort" on fuel development and core performance and separate "supplemental efforts" for each type of reactor. At the very least, such an approach shows the commonality of the separate projects, and also points out the costs of adopting parallel programs for two fast breeder types once a single large-scale program of basic fuels research is in operation.

Development of fuel elements is the most expensive part of the consolidated fast breeder program. A much higher level of fuel rod performance is required, over a much longer lifetime, for both the sodium- and gas-cooled projects. The total costs of developing the two fuel elements can be reduced from those for separate projects by eliminating the common expenses in fabrication and testing particular alloys in both; the sum total should not be greater than costs quoted above for the LMFBR fuel development plus those for gas as an incremental activity.

The results are going to be the same for one fuel core as for the other — the fuel rod and core development costs are going to be "target" for the gas fast breeder reactor if they are "target" for the liquid metal breeder master. The basic physics research will either provide leads for inexpensive development of a fuel alloy, for both or none. Then there are nine possible results from research of "high," "target," and "low" costs of fuel element and core development, given the division of possible costs into these categories and identical results at each of these two stages for gas and liquid metal environments.[26]

Component development, in contrast, is aimed at better equipment for control of the core or operation of particular coolant systems — such equipment as control rod drives, heat exchangers, pumps, seals, and valves. The expenses for this work are then related to individual types of reactors. Heat exchangers for sodium are different from those for gas fast breeder designs. There are exceptions — for example, concrete pressure vessels are potentially useful in designs for all concepts and might be a general project, but even here it is central in the gas designs and peripheral in designs for the other reactor type, so that its costs are attributed directly to the gas-cooled fast breeder. The costs for all components are separate for particular reactor experiments in the consolidated project, so that there are nine possible results of "high," "target," and "low" costs for both liquid metal and gas research at this stage.

[26] That is, if TH is the result for rod and core research on the LMFBR part of the joint project, then TH is the result on both parts. There are nine such permutations.

Prototype construction in any project does not differ when research is taking place on more than one type of reactor. A reactor experiment with a sodium heat exchanger cannot be used to learn very much about energy production from a plant circulating gas through the core. Development expenses are unique to each concept; there are again nine possible permutations of "high," "target," and "low" costs on liquid metal and gas prototypes.[27]

A broadly based research project to build two efficient fast breeders would start with the general or "mainstream effort" and add specific programs in the coolant concepts. The costs of such a program would vary widely. Given the range of costs estimated, there are $(9)^3 = 729$ permutations of "high," "target," and "low" costs at each of three

Table 2.8 Estimated Costs of a Single Project Producing Two Reactor Types

Probability of Costs Less Than Those Shown	Costs of Research (billions of dollars)
0.01	2.2
0.10	2.6
0.20	2.9
0.35	3.1
0.50	3.4
0.65	3.6
0.80	3.9
0.90	4.2
0.99	4.6

development stages which comprise a frequency distribution of possible results. The distribution is outlined in Table 2.8. It shows clearly that, whatever the results, the consolidated project for two reactor types is cheaper than two parallel and separate projects. When merged, the two cost 1.5 times the single liquid metal project, or 1.7 times the gas project. But the merged project promises two reactor types.

A variant on the second approach would be to start two reactor projects in a consolidated program, but to continue only the promising or "low cost" one to the end. This is not promising of lower total costs for one reactor — at least not at this stage of research. Not enough can be learned about fuel performance to choose the better of the two

[27] For example, "high" costs could be realized on the liquid metal prototypes and "target" costs on the gas, so that one of the nine results is HT. Others are HT, TH, TT, TL, etc.

before the heavy expenditures on element and core testing have been made for both reactor types. From there, whatever the results on fuel research, those on components and prototypes are independent so that no information is gained from the work on components on the lower of the cost levels on prototypes. In effect, a portfolio of two projects from which to choose has costs as given in Table 2.8; the costs of only one reactor type in the consolidated program are from 1.5 to 1.7 times those in one separate LMFBR or GCBR program.

In Review

After a survey of some of the outstanding problems in developing fast breeder reactors, the costs of solving the problems have been estimated for sodium- and helium-cooled fast reactor types. The gas-cooled project seems to promise the lower expenditures, for any of a wide range of successes and failures. At least, with reasonable research success on the fuel core and no more than the expected number of prototypes, the gas-cooled project will cost close to $2.0 billion, while the sodium-cooled project will be $200 million more. With poor results, the gas reactor will cost $2.8 billion to develop and the sodium reactor will cost $400 million more.

One research program has been considered which would build demonstration reactors for both sodium and helium systems. The fuel core would be developed as a common project for both reactor types, insofar as that would be possible; then separate components in each technology would lead to separate demonstration reactors. The costs of doing so are expected to be considerably less than the sum of those for two parallel projects — from $100 million to $1.4 billion less. The question for the following chapters is whether two reactor types are better than one, or whether the gains outweigh the costs of the added technology.

3

Initial Forecasts of the Benefits from Breeder Reactors

The construction of a large breeder reactor prototype is not so distant in the future that decisions on number and types of such plants can be avoided. The technology of fast breeder prototypes, at least as far as core configuration and coolant are concerned, is being decided now by means of the allocation of initial research funds among laboratories and experiments. These decisions have been based upon limited forecasts of the final results of the research and on the implications of these results for the preservation of stocks of natural uranium in this country. They have to be put into perspective by an initial assessment of the benefits from fast breeders.

The benefits from any one of the research projects of interest will have to be found in the pattern of adoption of breeders by electricity generating companies. A successful project by definition has to result in breeder reactors being built by private equipment manufacturing companies. The benefits are the economic gains from doing so — the consumers' and producers' surpluses from adding to electricity generating capacity by means of these *specific* reactor types.

These benefits are going to be spread throughout the energy sector of the United States economy in the years between the completion of the first demonstration breeder plant and the first decade of the twenty-first century. They can be estimated as equal to the difference between forecast demands and forecast costs for breeders, over and above non-breeders, in the nine power regions in this country, in each of the five-year periods during 1985–2005. Demands for any type of reactors are

$$Q = f(P, p, Y, N) \tag{1}$$

for capacity installed Q, an array of capital prices P charged the buyers (the electricity generating companies) for alternative nuclear and nonnuclear systems, an array of forecast fuel cycle costs p for alternative fueling systems, the total income Y of final electricity consumers, and the population N of final consumers. The differences between demands for breeders and nonbreeders is $(Q_B - Q_{NB})$, where Q_B and Q_{NB} are forecast for the common values of Y and N, and for the values of P, p appropriate for each type of plant in each power region.

The costs of production of breeder reactors are defined as C with

$$C = f(A, p, Q), \tag{2}$$

where A is an array of factor input prices other than fuel, p is the aforementioned fuel cycle costs and Q is the added capacity (dependent upon the technical relation or "production function" for capacity and factor input quantities which results from the breeder reactor research projects).

These two conditions do not complete the description of the parts of the economy affected by breeder reactors. There must be some relation between prices P and breeder costs C dependent upon the nature and extent of control of capacity supply by the producers of that capacity. Also, fuel cycle costs are determined in the energy sector. The demands for nuclear fuels q_D are defined as

$$q_D = f(Q, p) \tag{3}$$

with the total tonnage of fuel demanded derived from the demand for capacity Q in the reactor markets and from the fuel cycle costs. The supplies of nuclear fuel q_s are defined as

$$q_s = f(p, Z), \tag{4}$$

where p is defined above and Z stands for forecast reserve stocks of various uranium and plutonium compounds. The demands for new or additional quantities of fuel are equal to the supplies at the market clearing price p^*, which then determines costs of capacity C and capacity demands Q_B and Q_{NB}.

This system of four equations can be used to forecast benefits by finding values in the energy sector of P, p, Q_B, Q_{NB}, and C which are consistent with conditions "outside" the energy sector later in this century. The "outside" conditions are represented by Y, N, A, and Z; by first forecasting per capita income Y/N, population N, and reserves of nuclear resources Z, and then inserting these into this equation system, the future effects of breeder reactors can be simulated. The simulation finds estimates of nuclear breeder and nonbreeder capacity at prices consistent with energy demands in the economy. Then measurements of consumers' and producers' surpluses can be made for these levels of capacity and prices.

Each of the four equations is discussed in the following sections. The discussion is quite general: the characteristics of demands and costs for breeders and breeder fuels are outlined, but not stated in quantitative terms, for conditions likely to prevail in the last quarter of the century. The particular conditions are specified in the last sections of the chapter on, respectively, "the economy without breeders" and "the economy with breeders."

Forecasts of the Demand for Nuclear Reactors

The contracts for fast breeder reactors, to follow after the successful operation of demonstration plants, could conceivably account for all of the new commitments to nuclear capacity after 1985, and even some of those which would have gone to fossil fuel plants. But there is some basis for believing that this will not be the case — that commitments for breeders will be only for certain plants built to the largest possible scale for base load electricity generation in large cities. The choice of nuclear plants over fossil plants, of fast breeders over nonbreeders, and among types of fast breeders, will depend on relative prices and fuel costs, and on the objectives of the electricity generating companies. These are not known at the present time, and one or the other set of plausible assumptions makes one or the other conclusion on the demand for breeders more convincing.

The buyer of electricity generating facilities is a corporation of great complexity with objectives so diverse that no monocausal explanation of patterns of demand is appropriate. The equipment purchased in the past shows that there has been continued search for more efficient means to produce electricity. The new plant and equipment has brought about an annual rate of productivity improvement of approximately 5.5 per cent since 1900, "more than three times the rate of increase of productivity for the economy as a whole" and at the same time "production has increased at about twice the rate of increase in overall industrial production."[1] Yet the companies have departed from an exclusive interest in adding the largest and lowest cost plant. The nonbreeder nuclear reactors purchased in the early and mid-1960's, at the demonstration or postdemonstration stage, required net expenditures per kilowatt hour far greater than those in alternative fossil fueled facilities. The companies have not only purchased "high cost" advanced systems, but have engaged in outright research where the specific company research gains were not obvious and where the public explanation was couched in terms of "the good of the industry." The new systems have generally had greater risk of temporary suspension of operations or even of long-term failure than alternative nonresearch systems. This has been the case in some first-round nuclear reactors

[1] The Federal Power Commission, *National Power Survey* (Washington, D.C.: U.S. Government Printing Office, October, 1964), p. 10.

built by industry with government grants for research components, and in regional high voltage interchange systems where the result has been higher levels of system performance with higher risk of area-wide failure.

An explanation of these diverse patterns of behavior is not really possible, given that some part of it is company policy, and that companies that are regional monopolies have the opportunity to be quite arbitrary. But there may be a "central tendency" to follow a particular pattern — if electric generating companies act as if they are profit-making firms subject to severe restrictions on the decisions most greatly affecting such profits. Adding to the profits of the company $R - C$, with R receipts and C expenditures on all inputs for providing Q of capacity, is the basic justification for an investment policy, whether this is made to stockholders or lenders of capital funds. But state and local regulatory commissions and the Federal Power Commission set limits on profits per unit of capital investment K, so that $(R - C)/K \leq r$, with r the commission-allowed rate of return. Then purchases of equipment must add to profits $(R - C)$ without increasing $(R - C)/K$ beyond the allowed rate of return.

The substance of such a policy consists of minimizing costs, except where and when profits can be increased by not doing so. Consider costs $C = F(p_1, K, p_2, F)$ subject to $\lambda_1[Q - f(K, F)]$ for the generating company's production function[2] and $\lambda_2(r - (R - C)/K)$ for profit regulation. The minimum marginal costs of energy are $\partial C/\partial Q = p_1/(\partial Q/\partial K) = p_2/(\partial Q/\partial F)$, equal to the identical ratios of input prices to marginal products of those inputs, capital K and fuel F. But the closest that the regulated firm comes to these minimum marginal costs is the price-marginal product ratio above, plus the allowed profits on capital per unit of capital.[3] The greater the allowed rate of return, the greater the deviation from minimum costs.

The argument implies that there is a difference in input purchase patterns between a regulated electricity generating company and an other company not under a profit constraint. More is used of capital relative to fuel in the regulated firm;[4] in fact, the producing facilities are capital using and fuel saving, even if the total costs of generation are

[2] The production function $Q = Q(K, F)$ is the expression of technical-engineering limits on output Q with given inputs. Here Q is megawatts capacity.

[3] To minimize costs, set the partial differentials of $[C(p_1, K, p_2, F) + \lambda_1 Q_1(K, F) + \lambda_2(R - C/K - r)]$ equal to zero. This results in an expression for marginal costs in terms of $(\partial Q/\partial K)$, P_1, λ_1, λ_2, and r, that is additive for the marginal product and the rate of return times the additional capital $\partial K/\partial Q$ per unit of capacity Q.

[4] The ratio of marginal products $(\partial Q/\partial K)/(\partial Q/\partial F)$ is now equal to $p_1/p_2 + \lambda_2/(\lambda_2 - 1)(r/P_2)$. Given that this last term is positive, then $\partial Q/\partial K$ is relatively lower than in the absence of the constraint — more capital and less labor is used than before.

increased by changing the capital-fuel ratio. With equipment as the favored input, those components promising lower fuel costs in the future would be most favored of all.

The argument has implications for the demand for breeder reactors, as compared to that for nonbreeders, in the period after 1985. The accuracy of long-term forecasts can be improved by constructing a framework that takes account of these motivations of buyers of equipment. The approach in this framework will be to assume that the choice of reactor types will be similar to the present-day choice between reactors and fossil fuel systems. This current choice will be generalized by fitting functions to historical data. The resulting analytical statement is then used as the basis for forecasting nuclear breeder and nonbreeder production capacity after 1980.

Consider that total new nuclear and fossil fuel capacity Q, in megawatts installed in a given year, depends upon the forecast average annual expenditure per kilowatt on plant and equipment P, and the forecast average price of fuel p per kilowatt hour generated in the lifetime of that equipment. The expectation is that $Q = f(P, p)$ shows the characteristics of declining demand for capital inputs in production $\partial Q/\partial P < 0$, or that installed facilities are going to be more extensive where capital prices are lower.

The additions to capacity depend not only on costs to the buyer but also on the demands on the equipment buyer to produce electricity. Electricity consumption increases with the growth of population, with increased incomes of consumers, or as a result of increased purchases of durable consumer's goods which use this form of energy. Consumption is limited to some extent by higher prices for this output. Then the demand for capacity Q can be considered to depend on the long-run demand for electricity E where $E = f(P_e, Y/N, N)$ with P_e the price of electricity expected over the lifetime of the plant, Y/N the per capita real income, and N the projected population in each electricity market. This dependence of capacity Q on E is $Q = f(P, p, \Delta E)$ with ΔE the measured or forecast change in E each year following from price, income, and population changes.

Additions to generating capacity each year are required not only to meet expanded demands for electricity, but merely to maintain the stock of generating equipment. The amount of replacement capacity added each year is equal to the depreciation rate "α" times the stock of equipment "K" accumulated during previous years of electricity generation. The demand for capacity includes both Q and $Q^* = \alpha \cdot K$, since replacement is required as well as expansion. Then total demand $Q = f(P, p, \Delta E, \alpha K)$ where new capacity additions include both those for expansion and those for replacement of worn-out equipment.

The demand function $Q = f(P, p, \Delta E, \alpha K)$ can be fitted to the historical behavior of prices, electricity consumption, and capital stock. Each year in the post World War II period provides an "observation" of Q, P, p, E, K; and least-squares regressions of the functional relationship can be computed from the available sample of these observations. This procedure chooses the parametric form that minimizes the sum of squared deviations of observed Q from the functional value of Q as a dependent variable. Forecasts of future capacity are then made by inserting estimates of the future values of the determining or independent variables into the fitted demand function.

The share of future demand for capacity going to nuclear plants depends on the relative capital and fuel prices of this technology. Nuclear facilities should be installed when nuclear capital and fuel costs are less than those from fossil fuel plants with the same designated lifetime capacities, because the nuclear facilities both reduce costs and add to the rate base for determining allowed profits. But nuclear facilities are also expected to be installed when their costs are not lower than those for fossil-fueled facilities because of preference for more research expenditures embodied in new facilities. The frame of reference must allow for less than perfect substitutability between types of production facilities.

The straightforward approach is to calculate the rate of substitution of nuclear for coal or oil facilities as a result of changes in relative prices. Assume that the proportion of additional nuclear to total capacity Q_n/Q in any power region depends on the prices of nuclear equipment P_n and nuclear fuel p_n, and upon the prices of least-cost alternative fossil equipment P_f and fuel p_f. A change in these relative prices should shift some, but not all, new plants out of or into the nuclear class. The approximate extent of the shift can be determined from fitting the general demand relation $Q_n/Q = f(P_n, p_n, P_f, p_f)$ to the experience with nuclear capacity additions in the period 1960–1970. The computed equation specifies the nuclear portion of capacity demand, given that present motivations continue to dominate in the future; this provides a framework for forecasting nuclear facilities to be purchased in the period after introduction of breeders. Forecasts are obtained from inserting estimates of future prices into this equation and multiplying the resulting estimate of Q_n/Q by total capacity demand Q.

The initial forecast of the demand for breeder reactors does not follow this proposed analytical framework in complete detail, because sets of observations have not been generated for each specified variable. A shortened version of the model, in keeping with the information that is available, has three parts. The first considers the present demand for all capacity $Q = (P, Y, N)$ and makes projections of capacity installed in the 1980's, 1990's, and early 2000's from this demand function. The second is a forecast of the nuclear portion of this new capacity $Q_N/Q = f(P_n, p_n, P_f, p_f)$. The third is an assessment of the change in the nuclear portion brought about by the introduction of fast breeder reactors in those three decades.

Estimated Reactor Demand Functions

Present demand for megawatts of electrical capacity is obviously related to the prevailing conditions of factor prices and of final electricity demand. These conditions may not hold in the future, but the functional relation between capacity demand and prices may be roughly the same now as in the last 20 years of this century, so that the present demand function can be used with forecast future prices to forecast future additions to capacity. The measured demands are additions to thermal capacity in electricity generating systems within any five-year period. New plants added over this period are substitutes because they meet the same future demands for electricity where the future is some years hence (and trade-offs of present for future plants can be made when there is lead time in meeting these electricity demands). The additions to capacity in particular systems making up a "power pool" should also be roughly competitive; the factors affecting additions at one location in the pool are the same as at other locations, in the sense that trade-offs of more power capacity at one location for less at others can be used to eliminate any "net disadvantage" in building at some one place in the power region. Then the *market* for new capacity is roughly bounded by the geographical limits on power pools and by a five-year period. The relevant markets in the last few years roughly conform to the nine power regions of the Edison Electric Institute in the periods 1958–1962, 1963–1967, and 1968–1972.[5] The total megawatts of thermal capacity installed varied greatly from region to region and period to period: from 2,805 MW in Region 4 in 1958–1962, to 24,540 MW in Region 3 in 1968–1972.[6]

[5] The Edison Electric Institute power generating regions are as follows:
1. Maine, New Hampshire, Vermont, Massachusetts, Rhode Island, Connecticut.
2. New York, New Jersey, Pennsylvania.
3. Ohio, Indiana, Illinois, Michigan, Wisconsin.
4. Minnesota, Iowa, Missouri, North Dakota, South Dakota, Nebraska, Kansas.
5. Delaware, Maryland, District of Columbia, Virginia, West Virginia, North Carolina, South Carolina, Georgia, Florida.
6. Kentucky, Tennessee, Alabama, Mississippi.
7. Arkansas, Louisiana, Oklahoma, Texas.
8. Montana, Idaho, Wyoming, Colorado, New Mexico, Arizona, Utah, Nevada.
9. Washington, Oregon, California.

[6] The amounts of capacity installed in each region have been compiled from the megawatt capacity ratings of all new plants shown in that period and region by the Federal Power Commission records. The installations for the period 1968–1972 have been compiled from forecasts and new contracts shown in industry surveys of future demands or from extrapolating the experience in new contracts for delivery in 1968–1969 to the entire five-year period. This is outlined in detail in Appendix C.

The most direct explanation of the extensive variation in installed capacity lies in the behavior of demands for electricity output. If a certain technical capacity were needed to produce a kilowatt hour of electricity output, then the determinants of electricity demand would be those for capacity demand. Although this is not exactly the case, the determinants of electricity demand are close enough to those for capacity so that aggregate income of consumers in this market and the price at which the electricity is sold can be considered the independent variables in the capacity demand function.

Relating the nine regional levels of population, per capita income, and prices of electricity sold in each of the five-year periods to additions to capacity shows the effect of these factors. The equation $\log Q = \alpha + \beta \log P_e + \gamma \log (Y/N) + \delta \log N$ for capacity installations Q, price of electricity P_e, per capita income Y/N, and population N, when fitted to these data by least squares is

$$\log Q = -7.046 - 1.240 \log (P_e) + 0.855 \log (Y/N) + 0.914 \log (N).$$

Each of the explanatory variables is statistically significant (in that the ratio of computed coefficients to standard errors of coefficients is greater than 2.0) and the three variables together explain approximately 64.8 per cent of total variation in installations of electrical megawatts of capacity ($R^2 = 0.6479$).

This characterization of capacity demand of electricity companies is held quite generally: the size of the market — in income and population dimensions — if not the price of electricity are usually assumed to affect the demand for new capacity. Here the extent of the first two effects is limited, as is generally thought to be the case. Income and population elasticities are less than 1.0, so that any percentage change in one of these variables is not matched by an equal percentage change in capacity additions.[7] The price effect is somewhat more pronounced, since the price elasticity of demand is -1.24; at least this is greater than might be expected from the results of more detailed econometric analyses of electricity demand.[8] The overall impression, however, is that this relation between capacity and determinants of final electricity sales should continue to hold in the future. There is no analytical reason for

[7] The "elasticity" of demand is defined as the percentage change in megawatt additions divided by the percentage change in the independent variable. If, for example, the price elasticity equals $(\partial Q/Q)/(\partial P/P)$, then in the log Q equation this ratio is $\partial(\log Q)/\partial(\log P) = \beta$. The estimates of the other elasticities are the coefficients in the log regression equation; they are all less than 1 or greater than -1. Then the demands are, as a first approximation, inelastic.

[8] Cf. C. Kaysen and F. Fisher, *The Demand for Electricity in the United States* (Amsterdam: North Holland, 1962), where a persuasive case is made for inelastic demand with respect to price changes.

the various elasticities of demand to change, even though there are reasons to expect different electricity prices, per capita incomes, and regional populations.

Prices of electricity have declined in the past, and it can be expected that they will continue to decline during the rest of the century. The past experience is summarized by the equation $P/\text{WPI} = \alpha_1 + \alpha_2 T$, where T is annual trend and WPI is the Wholesale Price Index; least-squares regressions of deflated electricity price in each region 1945–1965 show that the value of α_2 was negative in each region and the rate of decline varied from one half per cent point per annum in Region 6 to one and one-half points in Region 4.[9] Electricity prices for 1985 and subsequent years can be assumed to be on the separate regression trend lines at the 1965 Wholesale Price Index, at $T = 41$ and subsequent values.

Per capita income should increase at a substantial rate. The United States has experienced additions to each person's income roughly equivalent to 2.4 per cent per annum in the last half-century and, with effective Federal policies to attain the potential full-employment output, this rate should be increased to close to 3 per cent in the remaining years of this century.[10] It is assumed that additions of this magnitude are made to the present per capita income in each region (so as to preserve the present relative distribution of income for the rest of the century).

The population in all power regions is assumed to increase from roughly 200 million at the present time to more than 266 million in 1985. These are forecasts of the U.S. Census Bureau,[11] which imply

[9] The regressions were fitted in log-linear form and forecasts were made as shown in Appendix C.

[10] Cf. J. Kendrick and R. Sato, "Factor Prices, Productivity, and Growth," *The American Economic Review*, Vol. LIII (December, 1963), p. 974, where the annual percentage rate of change of real output is 3.15, the annual percentage change in population is 0.76, so that the per capita real output increase is 2.39. This is not equal to maximum potential output, since these averages for the period 1919–1960 reflect the depression of the 1930's and subsequent smaller downturns in the late 1940's and 1950's. The equilibrium growth path may well follow the aggregate production function $Q = AK^{\beta}L^{\gamma}$ so that the percentage time changes in real per capita income $Q'/Q - L'/L = \beta(K'/K - L'/L) + A'/A$. The equilibrium path is that at which there is no net tendency for *a priori* savings to depart from *a priori* capital investment. With a fixed savings rate and distribution of income, this is equivalent to $Q'/Q = K'/K$ which results in the equilibrium $Q'/Q - L'/L = (A'/A)/(1 - \beta)$.

With β, the proportion of national income received in capital returns, roughly equal to 0.25, and the annual rate of technical progress A'/A not likely to be less than 2.25, the equilibrium growth path for per capita income should not be less than 3 per cent per annum.

[11] Cf. U.S. Bureau of the Census, *United States Population Projections*, Series B (Washington D.C.: U.S. Government Printing Office, July, 1964). These projections are for "medium assumptions" as to economic and technical progress. Cf. the discussion of these forecasts in Appendix C.

that Region 1 should have 5 per cent of the population in 1985, Region 2, 17 per cent, and the other regions from 3 to 9 should have 19, 7, 16, 6, 9, 4, and 14 per cent, respectively. This particular distribution of population would be considerably different from that for 1980; if this rate of change is assumed to continue over 1985–2005, while total population increases at rates shown by the long-term census forecasts for 1985, 1990, 1995, and 2000, then forecasts can be constructed for each of the nine power regions.

The projections imply substantial increases in capacity demand. This is because the regions likely to experience the greatest increases in total population are going to have the highest per capita incomes, with multiplicative effects implied by the fitted demand function.

Table 3.1 Five-Year Additions to Capacity in Nine Power Regions

Power Region	1,000 MW of Added Capacity 1985–1989	1990–1994	1995–1999	2000–2004
1	11.5	15.9	21.7	29.8
2	40.5	54.1	72.0	95.8
3	66.6	94.6	133.6	189.1
4	17.1	23.7	32.5	44.7
5	40.3	56.3	78.3	109.1
6	55.0	81.4	120.0	177.0
7	28.8	41.1	58.6	83.5
8	16.9	24.5	35.3	51.0
9	53.4	74.1	102.4	141.6

SOURCE: $\log Q = -7.046 - 1.240 \log (P) + 0.855 \log (Y/N) + 0.914 \log (N)$ for forecast values of P, Y/N, N in each of the nine power regions.

Table 3.1 lists these forecast additions by power region. The low rate of population and income growth in New England, and continued high electricity price, can be expected to retard additions to capacity in Region 1; but even here the amounts expected to be added are 50 per cent greater in 1985–1989 than in 1968–1972. The extremely high rate of population and income growth in the north-central region should increase the annual addition to capacity there seven times over by the end of the century. The expected increases in the other regions are substantial — so that the total added capacity between the years 1985 and 2000 is expected to be 1,450 thousand megawatts.[12]

[12] Cf. U.S. Atomic Energy Commission, *The 1967 Supplement to the 1962 Report to the President on Civilian Nuclear Power* (Washington, D.C.: U.S. Government Printing Office, 1967), pp. 18–19 for alternative Atomic Energy Commission and other agency forecasts.

What proportion of these additions will be new nuclear reactor plants? This is a matter of the relative prices of nuclear and fossil-fuel-fired plants and of the relative fuel costs for these alternative facilities. Consider the last decade in the history of nuclear plant contracts to show consistent buyers' reactions to relative prices; if the pattern continues, decisions to purchase or not to purchase more reactors in the 1980's and 1990's will depend on percentage changes in relative prices. The history conforms to the relative share demand schedule

$$\log(Q_n/Q) = \alpha + \beta \log(S) + \gamma \log(\text{PNF/PFF}) + \delta \log(\text{PNK/PFK}).$$

Nuclear capacity was installed in four of the nine power regions in the period 1958–1962, six of the power regions in 1963–1967, and eight of the power regions in 1968–1972. The average size of any plant S, whether nuclear or fossil, can be calculated from Federal Power Commission, Edison Electric Institute, and Atomic Energy Commission records. Capital and fuel prices for nuclear facilities (PNK and PNF) and for alternative fossil fuel facilities (PFK and PFF) were known to some extent by each of the purchasers before the installation of this new capacity — the nuclear prices from the plants that were constructed, the fossil fuel prices from actual expenditures on the last fossil fuel plant built in each power region.[13] Percentage additions to capacity accounted for by new nuclear plants are shown by the least-squares regression equation

$$\begin{aligned}\log(Q_n/Q) = &-4.375 + 0.418 \log S \\ &- 0.859 \log(\text{PNF/PFF}) - 1.030 \log(\text{PNK/PFK}).\end{aligned}$$

This is as expected from the purchase patterns of regulated companies.[14] The higher the relative price of nuclear fuel, the smaller the percentage of capacity additions that are nuclear plants. The important determinant of the nuclear percentage is the relative price of nuclear capital: the higher the relative price of nuclear capital, the lower the nuclear percentage of total demand, and any percentage difference between nuclear and fossil capital prices brings about a greater proportionate change in (Q_n/Q) than the same percentage difference in fuel prices.

[13] There are 13 observations in the nine power regions in the three time periods of additional nuclear megawatts of capacity, expected prices for this new capacity, and the last experienced prices for fossil fuel capacity. Nuclear fuel costs and alternative fossil fuel costs were not usually specified at the time the contract for the new plant was signed. But estimates of assumed cost can be made from the technical characteristics and assumed rates of operation of the new plants. These are shown in detail in Appendix C.

[14] The equation has reasonably satisfactory "goodness of fit" characteristics: the coefficients divided by estimated standard errors are +0.881, −2.722, −2.512, respectively, and the coefficient of multiple determination $R^2 = 0.818$. It should be strongly noted that there are only nine degrees of freedom, given the 13 observations of other than zero nuclear shares.

The demands for nuclear reactors in the 1980's and 1990's should not be different in kind from those in the last decade. Relative prices of capital and fuel should differ from those experienced, but there is no basis for expecting that the (Q_n/Q) function would have greatly different exponents from those shown for the earlier period. Adjustment to new technology should not be any more complete during the last two decades of this century, because fast breeder reactors will be at the same stage of introduction into existing fossil and light-water reactor systems as the light-water reactors have been in recent years. The "demonstration effect" of first-of-a-kind installation should have run its course for the light-water reactor, but would have just begun for the fast breeder. The favoritism of capital intensive projects, even at the expense of higher total costs of electricity, should have the same effect on the share for the more capital-intensive breeder as on the share for the water reactor over the fossil fuel plant.

Both nuclear capital prices and fuel costs should decrease so that favoritism toward nuclear capacity should be even more evident. This will be the case for nuclear versus fossil plant purchases, and also for light-water versus fast breeder installations. The lower relative prices should result in strong favoritism toward the new reactors. The actual forecast percentage shares, however, follow only after specific forecasts have been made of prices for each reactor type.

The Costs of Fast Breeder Reactors

The second determinant of fast breeder prices and quantities is the nature of costs for the manufacturer of providing that type of capacity. "Costs" are not "prices," which are expenditures of the buyers, but rather are the total outlays of equipment manufacturers required to put another reactor on the order line. They do not include research and development, or in most cases the long-run expenses for design and experimental production of improved systems. They are the design, components, and construction expenses for one more of those fast breeder reactors expected to follow soon after the demonstration plant.

There are two sets of parameters which are conditions for these "marginal plant production costs." The first are the technical limits on output of plant capacity from inputs of capital and labor — limits summed up by the "production function." The second are input prices.

The First Cost Determinant: The Production Functions of Fast Breeder Reactors

Reactors added to electricity generating systems in the 1980's will be governed by technical limits on output from various inputs that can be expressed as "production functions" $Q_n = f(K, F)$ for megawatts of capacity Q_n dependent on physical units of capital equipment K and on the inventory of nuclear fuel F. The limits are set by the state of the art for any type of reactor: a particular type, using materials with limited conductivity and durability, is capable of delivering only so much thermal energy from atomic fission to the turbine generator. Breeder reactors, as a new state of the art, will have new and different production functions in the sense that new combinations of capital and fuel will produce capacity. If liquid metal and gas-cooled fast breeders are both developed, then there will be two new and different production functions; it would then be possible to obtain capacity from a relatively larger number of combinations of capital and fuel, than now given by the light-water reactor technology.

Some of the additional combinations could be most interesting. One could turn out to be the least cost combination: with input prices expected to be P_K^* and P_F^*, amounts of capital and fuel K^* and F^* produce Q_n at least total expenditure, and a particular breeder type might attain K^*, F^* for Q_n while the other breeder and nonbreeder reactors would not. But capital and fuel prices are not known for certain. A type of breeder might be the only technology able to produce Q_n for a wide range of possibly least-cost K, F combinations. If one of these combinations does prove to be "least cost," that breeder would add to the economy by reducing the cost of resources used to generate electricity.

In the 1960's, rather than the 1980's, these combinations can only be investigated by forecasting the 1985 production functions. The forecasts may be wrong — excellent research results could reduce K for given F and Q_n, or research failures could result in higher values of K than called for in present design studies. But the forecasts can be made from some highly detailed engineering analyses of the 1985 breeders in a fashion which accounts for some of the possibilities of error. The "design studies," descriptive of plant operations as well as equipment, can be considered indicators of "central tendency" for Q_n, K, and F for 1985 plants because they direct the research now taking place. To a reasonably large extent, the fuels and components experiments are undertaken to achieve "targets" of performance set by the design studies; with research measures aimed at the design characteristics, the probability that they are going to be achieved must be large.

There is information in the design studies for estimating the parameters of production functions. All functions seem to be within the bounds of the general form

$$Q_n^{\delta} F_b^{\pi} = \alpha K^{\beta} F^{\psi}$$

with Q_n the maximum sustainable capacity for producing electricity during each output period over the forecast reactor lifetime, and F_b the bred fuel output from capital K and fuel F, subject to reductions or increases in magnitude as shown by the constant term α. This form is plausible because it implies that the marginal products of capital in either capacity ($\partial Q_n/\partial K$) or breeding ($\partial F_b/\partial K$) depend upon the amounts of the other input and of the other output (a pattern of marginal products which has generally been observed to follow from variations in output temperature and fuel rod design). Rewriting the relation so that

$$Q_n = \alpha K^{\beta/\delta} F^{(\psi-\pi)/\delta} = \alpha K^{\beta'} F^{\psi'}$$

— treating bred fuel as equivalent to input fuel — then information is required for estimating β' and ψ' for the different types of breeder reactors.

Useful information can be derived from performance analyses at the many various coolant temperatures and pressures tried in the design studies. These show the trade-offs of fuel for capital in the design, or $\partial K/\partial F$ for a given reactor size. This differential in the general form of the production function is

$$\partial K/\partial F = -\psi' K_1/\beta' F_1. \tag{5}$$

In addition, information is obtained from other engineering studies in the designs of the 1985 systems on the capital and fuel requirements for various sizes of a type of reactor. The studies of scale produce values of K_1, F_1 for Q_1 and K_2, F_2 for Q_2, for two sizes Q_1 and Q_2 of any reactor. These values can be inserted in the production function definition — so that $Q_1 = \alpha K_1^{\beta'} F_1^{\psi'}$ and $Q_2 = \alpha K_2^{\beta'} F_2^{\psi'}$ — and solving these for the unknown exponents β' and π' results in

$$\beta' \log (K_1/K_2) + \psi' \log (F_1/F_2) = \log (Q_1/Q_2). \tag{6}$$

Equations 5 and 6 provide sufficient detail to solve for β' and ψ'.

Estimates of β' and ψ' have been made from this type of information for each fast breeder type. They are based on this procedure, which can be called a "sensitivity" analysis for determining the input requirements for construction and operation of future breeder reactor capacity. They are not the result of deliberate tests of performance with more capital and less fuel, for example, but rather follow from comparisons of designs roughly similar in all respects other than capital/fuel ratios. As such, they provide a synopsis of today's views on technology for the 1980's and 1990's.

There are indications even in an outline of production functions that the technologies of the different breeder reactors are quite different.[15]

[15] Detail on the estimation of the production function is given in Appendix D.

This is shown both in individual observations of Q_n, K, and F, and in the effects of increasing inputs K, F on the scale of output.

The first estimates of α, β', ψ' are as follows. The liquid metal reactor production function is close to

$$Q_n = (2.5)(10^{-9})K^{2.9}F^{1.1}$$

in the range of capacities between 200 and 1,000 MW for reactor power ratings within 20 per cent of 1,000 kW/kg of plutonium fuel and reactor temperatures and pressures within 10 per cent of 1,000°F and 2,500 psia. There is reason to believe that this target capacity will not be attained at all times and the achieved capacity will be

$$Q_n = (2.5)(10^{-9})K^{2.9}F^{1.1}e^{-(\Delta T/T)(1-0.005)},$$

where $\Delta T/T$ is any transient temperature variant and -0.005 is the temperature-induced elasticity of reactivity. The gas reactor has the production function, under similar transient temperature variations, described by

$$Q_n = (6.7)(10^{-2})K^{1.5}F^{-0.04}e^{-(\Delta T/T)(1-0.008)}$$

over the same range of operating conditions.

In moving from small to large scales of operation, the liquid metal fast breeder requires smaller additions of capital and fuel than the gas reactors. If capacity is increased from 500 to 1,000 MW in a design liquid metal reactor, the accompanying marginal increase in capital and fuel is not more than 20 per cent of the amount for 500 MW. (Given that the necessary increase in inputs Δ is equal to $\Delta^{\beta'+\psi'} = 2.0$, and $\beta' + \psi' = 4.0$, then $\Delta = 1.19$.) The increase in requirements for the gas-cooled reactor is 65 per cent of the half-scale capital and fuel when moving to this higher level of capacity (if $\beta + \psi \sim 1.4$, then $\Delta = 1.65$). The economies of full scale accruing to liquid metal are larger than those expected in gas systems.

These are initial impressions. A study of the behavior of these production functions given expected ranges of factor prices can indicate the differences in technologies more clearly. The terms of reference are the costs of producing output — of additional capacity to produce electrical energy — from the different systems.

From Production Functions to Marginal Costs of Fast Breeder Capacity

The manufacturers of reactor equipment design and construct fast reactors that are described by these production functions. The sales involve costs for design and construction, and these costs are determined by the amount of new capacity from given amounts of capital and fuel. But the production functions are not the only determinants of costs for the reactor manufacturers. The levels of prices for capital components and for uranium and plutonium fuel have important effects upon the costs of this new capacity.

Once the technology and basic design for the type of reactor is available, a plant can be constructed at various sizes from 200 MW to more than 1,000 MW. The cost of adding one plant of any size to the manufacturer's output schedule is the "marginal cost of capacity." To minimize costs of providing capacity — where this depends not only on capital but also the present value of the stock of fuel — a design ratio of inputs has to be determined and then the scale of all inputs found. The decision depends both on factor prices and the dimensions of the production functions: total costs $C = P_K \cdot K + P_F \cdot F$ depend on capital and fuel prices P_K, P_F and the quantities K, F constrained by the production function $Q_n = \alpha K^{\beta'} F^{\psi'}$. The least-cost combination of K and F is given by the partial differentials of

$$C^* = P_K \cdot K + P_F \cdot F + \lambda(Q_n - \alpha K^{\beta'} F^{\psi'}),$$

where these partials are equal to zero; the level of minimum costs in keeping with this combination is defined as

$$C^{**} = \alpha^{-1/(\beta' + \psi')} Q_n^{1/(\beta + \psi')} [P_F ({P_F}^{\beta'}/{P_K}^{\psi'})^{-\beta'/(\beta' + \psi)} + P_K ({P_F}^{\beta}/{P_K}^{\psi'})^{\psi/(\beta' + \psi')}].$$

The marginal costs of providing thermal capacity in this technology are defined as C^{**} for the additional plant, or these costs per kilowatt are C^{**}/kW for the last plant on the production line.

The marginal costs of additional thermal capacity in an LMFBR in the 1980's should follow from the dimensions of the production function outlined in the previous pages and from factor prices forecast in the General Electric and Westinghouse design studies.[16] The production function values $\alpha = 2.5(10^{-9})$, $\beta' + \psi' = 4.0$ have been sketched out from the sensitivity analyses of fuel power and plant size.[17] The unit price of capital is taken as an index number $\sum P / \sum Q$ for all components from the same studies (from forecasts of nuclear engineers in these companies based, for the most part, on the present prices of components). The price per kilogram of fuel follows from adding fabrication, recovery, and uranium costs to the price per gram of plutonium where all of these are forecast. These costs include expenditures on the axial and radial blankets of uranium required for breeding around the plutonium core as well, so that total expenditures on all fertile and fissile materials are divided by the number of kilograms of plutonium to approximate a unit expenditure on fuel. The estimators

[16] As noted in Appendix D.
[17] Cf. the detailed discussion in Appendix D.

α, β', ψ', P_K, P_F inserted in the previous cost equation for C^{**}/kW show what it would cost to construct the General Electric type of LMFBR in the mid-1980's. The appropriate values of the same estimators — but different values from these for the LMFBR — can be used to find marginal costs for the gas fast breeder reactor. The gas reactor costs at 1,000 MW, the largest size within the range of the general design studies, are calculated by inserting $\alpha = 6.7(10^{-2})$, $\beta + \psi = 1.4$, the index of capital prices $\sum P/\sum Q$ for the manufacturer, and P_F for the lifetime fuel cycle costs. The prices $P_K = \sum P/\sum Q$ and P_F are the same as for the liquid metal reactor.[18]

These are forecast costs resulting from expected input factor prices and from the production functions implied by the major design studies. They may very well not be realized forecasts, given variations in either price or productivity conditions from those specified.[19] Reasonable or probable variations in prices would reduce these costs. Variations in operating performance could increase them very slightly.

The price of a unit of capital is based on implicit assumptions of productivity change in the industries supplying components between the mid-1960's and the mid-1980's.[20] A decline in components or material prices — in 1965 dollars — of one per cent per annum seems as likely as the negligible decline expected in the design studies. Then $P_K = 0.9 \sum P/\sum Q$ rather than $\sum P/\sum Q$, because of the "learning by fabricating" experience of suppliers of common components, similar to that experience most obvious in the early and middle history of the light-water reactor programs.

The forecast price of a unit of fissile fuel in the inventory specified for the production function depends directly on assumptions as to fabrication and recovery charges, and as to the prices per gram of uranium and plutonium in the 1985–1999 markets for these fuels. The fabrication and recovery charges vary from those implicit in the General Electric fuel prices because of technical progress in uranium and plutonium refining. The prices of plutonium and uranium compounds in the fuel core may well vary because of uncertain supplies of natural resources.

[18] The General Electric factor prices are used in estimating marginal costs for the gas breeder. This is appropriate because the units of capital and fuel have been scaled for compatibility with the LMFBR in estimation of the production function.

[19] The variations in productivity in the $Q_n = f(K, F)$ function are shown in the term $e^{-(\Delta T/T)(1-\eta)}$ for the transient temperature variations occurring in the day-to-day operations of the reactor.

[20] These forecasts are made in M. C. McNally *et al.*, *Liquid Metal Fast Breeder Reactor Design Study*, General Electric, Rept. GEAP-4418, Part 4 (January, 1964), Section 2.8.3. They were made on "a dollars/pound basis with the applicable unit cost varying with complexity and functional requirement . . ." (2–200) on what seem to be very conservative expectations for improvements in these supplying industries.

The Supplies and Demands for Uranium and Plutonium Fuels

The purchases of initial inventories of plutonium and uranium for liquid metal fast breeders will take place under supply–demand conditions quite different from those currently being experienced, and perhaps quite different from those envisioned for fueling nonbreeder reactors exclusively in the mid-1980's. This can be seen from a short review of different forecasts of supply and demand conditions.

The presumed supply conditions do not differ greatly from forecast to forecast. The country's known reserves of uranium, and the stocks of plutonium in the hands of the Atomic Energy Commission, will have to be drawn upon to provide new inventory. The rate at which additions and withdrawals of natural or manufactured reserves take place will depend upon the prices offered for fuel oxides or carbides by the equipment manufacturers or the electricity generating companies. National reserves have accumulated in the last few years as a result of a few significant discoveries; "in the period 1952–1958, U.S. uranium reserves have increased from approximately 4,000 tons to something in excess of 180,000 tons of U_3O_8... the large reserves of Ambrosia Lake in New Mexico, Big Indian Wash in Utah, and the Gas Hills and Shirley Basin in Wyoming were being developed."[21] An extrapolation of this early experience indicates that substantial additions will be made to reserves before the mid-1970's: "the number of discoveries that were made over a short period of time and with little exploration background strongly supports the position that there is much more uranium to be found in this country and that, given a market, it will be found."[22] Industry and Atomic Energy Commission estimates of supply — including those undiscovered but existing reserves — center on 500,000 tons of uranium oxide which could be produced and sold at a price of $10 per pound or less.[23] Further additions to supply can be had at higher prices. The assessments of reserves are generally based upon a direct relationship between greater expenditures on exploration and development in prospect of high uranium prices and additions to the supply of uranium oxide. From reserves plotted against price, the relationship $\log Z = a + bP$, where Z is tons of uranium oxide, and P is the price of this oxide per pound, seems most descriptive with $\log Z = 11.3 + 0.16P$ when only tonnage from "proved reserves" or "quite likely reserves"

[21] Atomic Industrial Forum, *U.S. Uranium Reserves: A Report by the Committee on Mining and Milling* (August, 1966), p. 2.

[22] *Ibid.*

[23] *Ibid.*, where 425,000 tons are expected to be available; and Atomic Energy Commission, *Civilian Nuclear Power Supplement*, p. 6, where 525,000 tons are "reasonably assured" and "estimated additional" for less than $10 per pound U_3O_8.

is included. Optimistic forecasts call for supplies of $\log Z = 12.5 + 0.16P$ or, at $10 per pound, of one million tons.[24]

There is less agreement among the many forecasts of the supply of plutonium fuel and of the extent of demands for the two fuels. Both depend on forecasts of the demands for electricity in the last quarter of this century, and on the share of generating capacity in alternative fossil- and nuclear-fueled systems. There are a number of such forecasts, some of which differ by one-half on the amount of plutonium produced in nonbreeder reactors, and on the capacity of nonbreeders which will be installed and thus able to produce this fuel.[25] Each megawatt of reactor capacity, given present patterns of in-core fission and energy conversion, could produce from 0.15 to more than 0.4 kg of plutonium per year: the annual additions to stocks of plutonium $Zp = f(Q_n) = 0.15 \sum Q_n$ to $0.40 \sum Q_n$ where $\sum Q_n$ is cumulative installed nonbreeder capacity.[26] By the mid-1980's there could be between 150 and 800 metric tons of plutonium from accumulation over the previous 15 years.

The demands for nuclear fuel are derived from final demand in the markets for electricity and from the costs of alternative fuel systems. The greater the utilization of electricity in the last quarter of the century, the greater the demands for nuclear fuel. But these demands also depend upon fossil and hydroelectric energy as alternative means for producing this electricity, and thus ultimately on the relative prices of the nuclear fuel and capital themselves.

A very simple formal model can be outlined for the interactions between fuel prices and final demand prices for electricity. Extensive additional research has to be completed before this model of the two interrelated markets can be used for forecasting purposes, however. But the model might provide a framework for further work; assume that the quantity F of nuclear fuel available in any one year is

$$F = a + bP_F + cZ$$

[24] These estimates are implicit in S. Golan *et al.*, "Uranium Utilization Patterns in a Free and Expanding Nuclear Economy Based on Sodium-Cooled Reactors" (presented at an Atomic Industrial Forum Conference, November 27, 1962), Fig. 5, Uranium Reserves and Resources in the U.S.A., p. 11. But they are consistent with — even quite close to — the Atomic Energy Commission and Atomic Industrial Forum estimates.

[25] The projections of installed capacity range from 80,000 to 110,000 MW_e in the Atomic Energy Commission, *Civilian Nuclear Power Supplement*, to more than 125,000 MW_e by the General Electric Company (McNally *et al.*, GEAP-4418, p. 14). The two numbers bracket the Edison Electric Institute forecast of 109,000 to 117,000 MW_e but exclude the Westinghouse forecast of more than 150,000 MW_e for total installed nuclear capacity (*ibid.*). This last forecast seems less probable, particularly in the absence of theoretical apparatus to support it.

[26] Cf. Golan *et al.*, "Uranium Utilization Patterns," and Edison Electric Institute, *Fast Breeder Reactor Report* (1968), p. 17.

where P_F is the price offered for newly mined output and Z is the stock of known reserves. The aggregate demand for these $F^* = f(P_F, P_K, R)$ as determined by the prices of fuel and capital and the revenues R from sale of the capacity made available by the fuel. The demand price is equal to the value

$$(\partial R/\partial F) = (\partial R/\partial \mathrm{MW})(\partial \mathrm{MW}/\partial F)$$

from the purchase of these supplies; when $\partial R/\partial F = P_F$, the reactor purchasers minimize capacity costs and set optimal (profit) levels of nuclear capacity. Then

$$P_F = \frac{F - a - cZ}{b}$$

and $P_F = (\partial R/\partial \mathrm{MW})(\partial \mathrm{MW}/\partial F)$ for each buyer and $P_F = n(\partial R/\partial \mathrm{MW}) \times (\partial \mathrm{MW}/\partial F)$ for n homogeneous buyers. It follows that

$$\frac{F - a - cZ}{b} = n(\partial R/\partial \mathrm{MW})(\partial \mathrm{MW}/\partial F).$$

With $\partial \mathrm{MW}/\partial F = \psi \mathrm{MW}/F$ from the production function, $F(F - a - cZ) = b\psi \mathrm{MW} n(\partial R/\partial \mathrm{MW})$ and equilibrium fuel consumption can be determined from the forecasts of (a) reserves, (b) the production function, and (c) final electricity revenues equal to $n(\partial R/\partial \mathrm{MW})$ for the n electricity producing power pools of roughly equal size. The equilibrium fuel consumption can then be used to forecast price P_F from Equations 1 or 2.

The approach that has to be taken here, given the preliminary nature of data available at this time, is simpler than this. The substitution of other forms of energy for electricity resulting from high fuel prices is ignored. The substitution of other systems for reactors is assumed to be determined by the demands for capacity $Q_n = f(\mathrm{PNK}, \mathrm{PNF})$. Thus the demands for fuel over a reactor lifetime are $F = I \cdot B \cdot Q_n$, the projected fuel/output ratio I in kilograms per megawatt year of capacity Q_n, times B (the present value factor for lifetime inventory requirements), times Q_n capacity. The values of I and Q_n vary with the type of reactor, the first being a technical constraint and the second endogenous to the system of supply–demand equations. Different fast breeder reactor types require different amounts of plutonium, as has been shown in the basic designs for the 1,000 $\mathrm{MW_e}$ plants; the present value of the stock of fuel in LMFBR is 558 kg for 1,000 $\mathrm{MW_e}$, but for the gas-cooled breeder the stock is -350 kg. The nonbreeder reactors have much greater lifetime fuel/output or $I \cdot B$ ratios, varying from 3,600 kg to more than 4,500 kg, depending on the development of technology to recycle plutonium as a substitute for uranium, and on the fuel cycle efficiency of particular nonbreeder systems.

Supply in this simple model of nuclear fuel markets then consists of stocks of uranium and plutonium that can be mined or produced, and then refined, at various prices for use in the reactor systems. The supplies are equivalent to the expected demands at some forecast prices. The demands are lifetime stocks needed to fuel forecast thermal capacity and are taken to be, as a first estimate, in fixed proportion to that capacity.

Nuclear Prices and Quantities in Markets Without Breeder Reactors

Energy production in the last quarter of the century might possibly be dominated by nonbreeder or thermal reactors. There might be no breeder development program. The 1970–1975 round of construction of full-scale commercial nonbreeder reactors could be followed by contracts for all the 1,000 MW plants to be built in the 1980's and 1990's — plants promising higher operating rates and lower construction costs, as custom production is replaced by assembly-line preparation of designs and components. But then again this may not be the case. The prices in uranium fuel and reactor capital markets both might be prohibitive.

An initial forecast of nonbreeder adoptions and prices is necessary in order to assess the effects of introducing one or two types of breeder reactors after 1985. Additions to reactor systems, and prices of capital and fuel, when there are only thermal reactors, are the reference points from which to measure *net capacity additions* and *price reductions associated with the new technology*.

The forecast is made by solving the four-equation supply and demand system for nonbreeder circumstances. The demand for reactors $Q(Q_n/Q)$ can be evaluated for exogenous population, per capita income, prices of fossil fuel and capital. The supply prices of nonbreeders can be assumed to be those presently quoted, but with adjustments for productivity changes likely to occur before 1985. The supplies and demands for uranium and plutonium fuels can be assumed to be in keeping with sales in this country of light-water reactors until the end of the century.

The Nonbreeder World: Demands and Prices of Capacity

First, the total demands for capacity are assumed to be as shown in Table 3.1. There is no reason to believe that "the type of the reactor" feeds back into the total demand for capacity, so that $Q = f(P_e, Y/N, N)$ is as outlined above for the nine power regions between 1985 and 2005. The nuclear share Q_n/Q, however, depends upon the average size of nonbreeder and fossil fuel plants, and on their respective prices.

The sizes of the pieces of equipment needed in each system at each point in time has been forecast, at least so that an initial assessment of Q_n/Q can be made. The Federal Power Commission's *National Power Survey* showed continued demands in the 1970's and early 1980's for small generating plants to meet peak loads and to provide safety margins against outage of 1,000 MW_e plants. But the relative importance of small plants should decline, so that by 1980 only 29 per cent of the generating capacity in the country will be provided by plants with ratings of less than 200 MW and only 14 per cent will be provided by plants between 200 and 400 MW. It is expected that plants with ratings between 400 and 800 MW will provide 11 per cent of capacity, and that the largest plants will be the most important of all, those with a 1,000 MW rating accounting for 45 per cent of capacity at that time.[27] If this size distribution of plants represents the results from minimizing expenditures on high-priced small plants — subject to safety requirements — then additions after 1980 should maintain these proportions. The producer of new reactor equipment could offer prices for plants in each of these size classes.

The prices of light-water reactors will be set by the two major firms now producing them — they now sell the dominant shares of new capacity, and it cannot be expected to be otherwise.[28] But the General Electric and Westinghouse Companies will not sell in the 1980's and 1990's at presently established prices. There is some reason to expect that there will be further development of these systems, or that the present working plans and production costs will not be frozen for the rest of the century. For one thing, "improvements by repetition" in the production of components should produce a three to five per cent reduction in costs each year, which should be reflected in at least a one per cent discount of present prices.[29] The continuation of present ways of pricing and market sharing ought to result in reactors of less than 500 MW being available at $315 to $320 per kilowatt in the mid-1980's. Similarly, reactors from 500 to 1,000 MW should sell at $135 per kilowatt, those with 1,000 MW of capacity or more should center roughly on $112 per kilowatt.[30] Twenty years of additional development experience — as reflected in lower costs and thus lower prices —

[27] Cf. Federal Power Commission, *National Power Survey*, p. 206.

[28] For discussion and analysis of the present controls over reactor supply, see Appendix A.

[29] A detailed review of cost indicators on all components has led to a forecast of a price reduction of only 0.5 per cent per annum. Cf. Jackson and Moreland, *Current Status of Light Water Reactors* (1968).

[30] This is based on twice the implicit price reduction of Jackson and Moreland (*ibid.*) on an annual rate of 1.0 per cent rather than their 0.5 per cent. Then this will lead to a not "unreasonable" pessimistic forecast.

might make the forecast average price for the size distribution of reactors less than this. But a first forecast has to take a doubtful view of prices less than $112 per kilowatt for the full-sized reactor in 1985.

The other systems obviously able to provide alternatives to new breeders are steam boilers fired by coal, oil, or natural gas. There is some evidence that price reductions will occur for these nonreactor electricity generating plants as well. There have been extensive rate reductions in the last decade on the transportation of coal in large volume,[31] and — since these still leave considerable disparity between rates and transport costs[32] — there should be another round of reductions of the same magnitude generated by the "competition" of more reactors. The results should be fuel cycle cost reductions of 0.5 to 1.0 mills/kWh in all power regions except 7 and 8.[33]

Further rounds of capital price reductions on fossil plants cannot be expected to be of magnitude equal to the fuel cycle cost reductions. The largest and newest coal-fired plants now show costs per kilowatt of installed capacity roughly equal to those associated with the plants built in the early and mid-1960's (after taking account of general price inflation). More pertinent, there is no incremental developmental program promising substantial productivity increases in the near future. The best present estimate of prices is $217 per kilowatt for the smallest size class, $153 per kilowatt for plants between 500 and 1,000 MW, and $133 per kilowatt for the full-size 1,000 MW plant in most parts of the United States. The least that can be expected is no annual price reductions, and the most is one per cent annual reductions.[34]

With this set of forecast capital and fuel prices, light-water reactors could expect to capture some part of the demand for additional capacity after 1985. The choice between these plants of various sizes might favor all fossil plants up to the largest size class, but potentially lower fuel cycle costs of light-water reactors might counteract this favoritism.

[31] Cf. P. W. MacAvoy and James Sloss, *Regulation of Transport Innovation: The I.C.C. and Unit Coal Trains to the East Coast* (New York: Random House, 1967).

[32] *Ibid.*, cf. for example, Table 3, p. 44, for marginal costs of $1.783 per ton and p. 133 for a rate of $4.12 per ton on 1962 trainload transport from the Clearfield coal district to the generating plant at Eddystone, Pennsylvania.

[33] Rate reductions which narrow profit margins for the railroads could be of almost any size. Assuming that the reductions are of the same order of magnitude as on the last round, however, profit margins would fall from roughly 100 per cent to roughly 50 per cent of calculated marginal costs on transport into the large generating stations on the eastern seaboard. The same reductions elsewhere — achieved on the first leg of delivery by barge or large volume ship transport — would bring about cost savings of from four cents to more than ten cents per million Btu, which are equivalent to 0.5 mills/kWh at the least and more than 1.0 mills/kWh at the most. Cf. Appendix E, where these are translated into annual fuel cycle cost reductions of 2.0 per cent.

[34] Cf. the detailed discussion in Appendix E.

Nonbreeder Reactors: Uranium Fuel Markets and Total Demands for Nuclear Capacity

The demands for nonbreeder reactors have to be forecast for a world in which these are the only types available. All determinants of that demand $Q(Q_n/Q)$ have been specified except the price of nuclear fuel PNF. This price is going to be set according to the supply and demand for U_3O_8 — at least according to the model.

The supplies of uranium can be "large" or "small," depending on whether further discoveries in the three untapped but potentially productive regions establish the supply function ($\log Z = 11.3 + 0.16P$) or the more expansive function ($\log Z = 12.5 + 0.16P$). The demands depend critically on uranium refining technology. With the knowledge of refining methods now at hand or expected to be available in the near future, the lifetime requirements of a reactor are going to be large. The core has to be removed every two to four years, the plutonium refined and stored for other uses, and a new core provided. Given the size of the average light-water reactor loading, the number of loadings, and a discount rate of 10 per cent, four tons of uranium fluoride will be required for every megawatt of capacity used over a 30-year period. But with new technology allowing the commercial recycling of the produced plutonium in *advanced nonbreeder or converter reactors*, the stock of lifetime demands is considerably reduced; self-produced plutonium should replace uranium on a four-to-one basis, so as to reduce inventory requirements to close to 2.5 tons per megawatt of capacity. Further improvements in the technology of refining can reduce the amount of uranium oxide even more (or, to an equivalent degree,

Table 3.2 Demands for Nonbreeder Reactors and Fuel Cycle Costs

Period	Additional Nuclear Capacity (10^3 MW)	Average Price ($/kW)	Additional Demand for U_3O_8 (10^3 tons of uranium)	Forecast Price of U_3O_8 ($/lb)	Equivalent Fuel Cycle Costs (mills/kWh)
1985–1989	59	127	237	8.9	1.3
1990–1994	72	122	289	12.8	1.5
1995–1999	93	115	370	15.7	1.7
2000–2004	121	109	484	18.2	1.8

SOURCE: Estimates from simulation with the four-equation model of reactor capacity and fuel quantities and prices. The values of the exogenous variables and the parameters are specified in the text.

the supply price of this metal). The low demands will most likely be close to two tons per megawatt capacity.

Then there are low estimates of supply and high estimates of demand, which comprise the basis for a "high" forecast of uranium prices. The more expansive supply forecast, along with the reduced demands from recycling and improved fuel processing, lead to a "low" forecast of uranium prices.

The "high" forecast price clears the uranium and reactor demand markets as shown in Table 3.2. The total demands for new, nonbreeder capacity in the nine power regions are expected to be close to 224,000 MW in the last 15 years of this century. The demands in the 2000–2004 period are expected to be relatively larger and to exceed 121,000 MW. In all, however, these do not add up to an optimistic forecast: most estimators look to total new contract capacity between 1980 and 2000 of 600,000 to 800,000 MW, while this set of five-year projections sums to 346,000 MW between 1985 and 2005.[35]

This forecast is relatively low because of the effects of "high" nuclear fuel prices and "low" fossil fuel prices on the nuclear share of total new capacity after 1985. The uranium fuel market, with limited stocks and requirements of 4 tons per megawatt of new capacity, is expected to clear at prices rising from $9 per pound in 1985 to as high as $18 per pound after the turn of the century. At the same time, fossil fuel price decreases of roughly two per cent per annum are anticipated — reducing this set of regional prices by 40 per cent in the last five years of the forecast period. These uranium and fossil fuel prices together have the effect of reducing the nuclear share, from close to 16 per cent of all new capacity in large or small plants in 1985–1989 to 11.9 per cent in the first five years of the next century.

The forecast nuclear capacity by region is found by multiplying total expected capacity Q by Q_n/Q for the relevant light water–fossil prices in each region. A tabulation of these shows the effects of relative fossil-nuclear prices in different parts of the country. Towards the end of the century, reactors should make up more than 15 per cent of the additional capacity in Regions 1, 4, and 9, and should account for more than 10 per cent of the additions made in all of the other regions. But nowhere can the light-water plants alone expect to make up more than 18 per cent of the total new installations of plants of all sizes (as shown in Table 3.3).

[35] Cf. Golan *et al.*, "Uranium Utilization Patterns"; The Edison Electric Institute, *Fast Breeder Reactor Report*; and Atomic Energy Commission, *Civilian Nuclear Power Supplement*.

Table 3.3 Capacity Accounted for by Nonbreeder Reactors, 1985–2004

Power Region	1985–1989	1990–1994	1995–1999	2000–2004
1	2.8*	3.3	4.2	5.4
2	7.9	9.2	11.2	13.9
3	10.6	13.1	16.9	22.2
4	3.3	4.0	5.0	6.4
5	6.1	7.5	9.5	12.4
6	8.0	10.4	13.9	19.2
7	4.6	5.8	7.5	10.0
8	2.8	3.6	4.7	6.3
9	12.9	15.5	19.6	25.2

* In thousands of megawatts.

SOURCE: As described in the text and in Appendix E.

The adoption of nonbreeder reactors could be much more widespread than shown here. With high forecasts for the prices of fossil fuels in each region, the nuclear share should be greater; with increased stocks and reduced demands of uranium from substitution of recycled plutonium, the nuclear share would not be cut off by high uranium fuel prices. Reasonable, if not somewhat optimistic, numerical values have been assigned to these assumptions to serve as the basis for this "high" demand forecast for 1985–2004. There is, in this case, no expected decrease in fossil fuel prices (so that, whatever the general price level in 1985, oil and coal prices stay at 1965–1970 relative price levels). But the new technology of recycling reduces the demand for uranium reserves from 4.0 tons to 2.0 tons per megawatt electric power, while the uranium supply function shifts out from $(11.3 + 0.16P)$ to $(12.5 + 0.16P)$ in keeping with ample discoveries in now-unexplored but potentially ore-producing regions of the United States.

Under these conditions, equilibrium in the fuel market and a continuation of present pricing in the equipment market should result in 555,000 MW of new nuclear capacity in the period 1985–2005. This is 200,000 MW more than expected from high uranium but low fossil prices. The details of this expansive forecast are shown in Tables 3.4 and 3.5.

The equilibrium uranium prices are very low — close to nine dollars a pound at most — and they imply reactor fuel cycle costs with recycling and advanced fabrication of only 1.3 mills/kWh. The range of

Table 3.4 High Demands for Nonbreeder Reactors and Fuel Cycle Costs

Period	Additional Nuclear Capacity (10^3 MW)	Average Price ($/kW)	Additional Demand for U_3O_8 (10^3 tons of uranium)	Forecast Price of U_3O_8 ($/lb)	Equivalent Fuel Cycle Costs (mills/kWh)
1985–1989	103	127	206	0.8	0.8
1990–1994	118	122	237	4.4	1.0
1995–1999	147	115	293	7.1	1.2
2000–2004	187	109	375	9.4	1.3

SOURCE: Estimates from simulation of 1985–2004 economic conditions with the four-equation model of the energy sector described in the text.

fuel cycle costs, from a low of 0.8 mills to a high of 1.8 mills, under the reasonably optimistic supply conditions implying Table 3.4, should bracket the costs most likely to be realized toward the end of the century.

The assumed capital prices are the result of changes likely to be made by the two large equipment producers in that period. The prices in the four size classes average $151 per kilowatt without accounting for the effects of cost reductions in equipment manufacture. With these cost reductions passed through in part as one per cent price reductions per annum, the average prices are as shown in Table 3.4,

Table 3.5 Capacity Accounted for by Nonbreeder Reactors Reactors Under "High Demand" Conditions, 1985–2004

Power Region	1985–1989	1990–1994	1995–1999	2000–2004
1	4.9*	5.5	6.7	8.4
2	13.8	15.1	17.8	21.5
3	18.4	21.4	26.7	34.5
4	5.8	6.6	8.0	10.0
5	10.7	12.3	15.1	19.1
6	14.0	17.0	22.1	29.7
7	8.0	9.5	11.9	15.5
8	4.9	5.9	7.5	9.8
9	22.3	25.5	31.0	39.1

* In thousands of megawatts.

SOURCE: As described in the text and in Appendix E.

beginning with \$127 per kilowatt (with \$110 for a reactor in the 1,000 MW size class) and ending with \$109 per kilowatt (with \$95 per kilowatt for the 1,000 MW reactor).

The sum total effects on regional sales of new reactors, given these prices of fuel and capital, are shown in Table 3.5. The concentration of sales in the East Coast population centers (Regions 2 and 3) and in California (Region 9) is apparent. Most of the additional 200,000 MW — over and above the commitments forecast under conditions of high reactor fuel costs — are obtained not by capturing all of the limited demands, but by taking 30 per cent rather than 15 per cent of very large total capacity demands in the large population centers. Here the fuel cycle costs of fossil systems are expected to be from 1 to 2 mills/kWh; the ability to undercut these costs given "expansive" nuclear fuel supplies, while incurring only slightly higher capital costs, can give nuclear reactors substantial shares. But, in a world of nonbreeder reactors, even the most expansive forecast implies minority nuclear shares of total new capacity in each power region.

Nuclear Prices and Quantities in Markets with Breeder Reactors

The introduction of fast breeders should increase the sales of reactors over those for the nonbreeder world. The promised lower fuel cycle costs of plants with the new technology provide an important and substantial basis for increased nuclear demand. The decrease in reactor prices from the entry and establishment of more producers would provide another, but not as substantial, reason for an increase in relative sales of nuclear-fueled plants. But the two reasons together can be expected to reduce the construction of nonbreeder reactors, replacing them with breeders, and also to replace fossil fuel plants with breeders so as to take the nuclear share closer to the total forecast additions to capacity after 1985.

This increased demand shows one dimension of the benefits from the breeder for the electricity generating companies and their consumers. The full measure is shown by the difference between demands up to capacity Q_0 for the light-water reactor systems in a market without breeders, and Q_1^* for breeder reactors and thermal reactor systems together (see Chapter 1). The benefits to buyers of generating systems are shown by the area between the two demand curves for these two alternatives. The area under the first demand curve can be estimated by finding $Q = f(P, Y/N, N)$ times $Q_n/Q = f(S, \mathrm{PNK/PFK}, \mathrm{PNF/PFF})$ for all prices over the range relevant to equilibrium in fuel and capital markets with nonbreeder reactors. The area $\Delta P \cdot Q$ can then be approximated for each difference in prices and summed over the entire range of relevant prices down to the equilibrium price P_0. The same area can be measured for the additional capacity accounted for by breeder reactors by assuming the same demand curves but the appropriate breeder prices.[36] The consumers' benefits equal the difference between the two areas for prices greater than or equal to the actual price that is forecast for that power region and time period. The benefits to producers are these prices minus the marginal costs of providing the additional capacity.[37]

The Demand and Cost Functions of Breeder Reactors

The introduction of breeder reactors into the established markets for fossil-fired and thermal reactor systems could have important and widespread effects on prices per kilowatt of any new capacity. There can also be large shifts of purchases from the older technologies to the new, even without price changes. Both price-induced and technology-induced reallocation of demands can be said to depend upon the demand function $Q(Q_n/Q)$ and the costs of producing reactors $MC^{**} = f(P_F, P_K)$, as a first approximation.

The demand schedules may very well differ from those for new additions to capacity in the 1960's, if the performance characteristics of the newer reactor types differ from those of nonbreeder reactors. This is going to be true to a limited extent — fuel longevity is expected to be greater, so that planned or scheduled "down time" should be less — but there is no basis for believing that the price and size elasticities in the share demand function Q_n/Q will change substantially. There will be major problems of unplanned "down time" in the breeder reactors and of operation of new technologies as part of a network of fossil-fueled plants; but these were encountered in the introduction of new nonbreeders and they determined to a great extent the computed coefficients of the Q_n/Q share relationship. The product of Q and Q_n/Q will be used to forecast the demands for breeders,

[36] The calculations are made by integrating the demand curves D_T for the light water reactor and D_B for the breeder reactor between Q_0 and Q_1^*. These demand curves are the same simple two-part demand functions $Q = f(P, Y/N, N)$ and $Q_n/Q = f(S, \mathrm{PNK/PFK}, \mathrm{PNF/PFF})$ for the different fuel and capital prices for nonbreeder and breeder reactor types.

[37] These marginal costs are for the given size distribution of plants for total amounts determined by the regional capacity demand. For Power Region 1 in 1985–1989, the forecast Q_n MW are divided so that A per cent of the nuclear generating capacity is in plants with less than 200 MW, B per cent of capacity in plants of 200–300 MW, C per cent of capacity in plants of 300–500 MW, and D per cent of capacity in plants of 1,000 MW. Marginal cost curves for each such plant size are summed horizontally for the marginal cost of providing total amounts equal to these percentages times Q_n MW. This is the case for all regional additions to capacity for the regional size distribution of nuclear plants.

whether of the liquid metal or gas-cooled type (or some combination), for the relevant prices of nuclear capital PNK and nuclear fuel PNF and the previously forecast fossil capital and fuel prices.[38] Then the residual nonbreeder share will be found by multiplying these estimated demands by the share equation Q_n/Q again, for the relevant non-breeder prices PNK, PNF against the breeder prices.

The marginal costs of producing reactor capacity are the lower bounds on predicted prices, just as the demand functions are the upper bounds. These costs are defined above as

$$\text{MC}^{**} = f(\alpha, \beta', \psi', P_F, P_K),$$

where α, β', ψ' are the parameters of the production functions of either liquid metal or gas breeder reactors and P_F, P_K are the prices that reactor manufacturers have to pay for units of capital and fuel used in building and then "providing" lifetime capacity. The production functions have been described, based on sensitivity analyses in the manufacturers' preliminary design studies of the two technologies, and the forecast index price of a unit of capital in the steam and generator system is \$16(10^4) from the same source.[39] The fuel price paid by the manufacturer is assumed to be set by the supplies and demands for plutonium in the core and uranium in the blanket of the new reactor (as for the operation of these reactor systems after they have been constructed).[40] Thus costs, and prices of reactors, are set simultaneously with equilibrium in the uranium and plutonium markets.

The Prices of Breeder Reactors

The quotations of manufacturers to the generating companies in any region for delivered capacity Q_n could very well range from the marginal costs on one end to many times these costs on the other. But the structure of the market — the uniqueness of these new reactor systems as shown in $Q(Q_n/Q)$, and the number and relative size of producers of these systems — should set the most likely price-cost

[38] The size variable S in $Q_n/Q = f(S, \text{PNK/PFK}, \text{PNF/PFF})$ is given the value for the median of each of the four size classes described above. Then the share Q_n/Q is calculated for each size class.

[39] The index of prices $\sum P/\sum Q$ = \$100,000 for the reactor system, and \$160,000 for reactor and turbine generator, from the information analyzed in Appendix C.

[40] That is, the reactor manufacturer "passes through" the purchase price of plutonium and uranium to the generating company without an additional markup, so that P_F in the cost function sets the fuel cycle costs PNF in the nuclear share demand function. This is to assume that "turnkey" construction with fuel costs guaranteed by the builder has the same price as contract construction and independent operation by the buyer.

margin. Also, the producers of breeders are going to be either new companies with new types of reactors, or old companies producing both the currently available nonbreeder reactors and the new types of reactors. The alternatives should have contrasting effects on the forecast price-cost margin.

The structure of the reactor capacity market can be described in terms of two historical "givens." The first has already been shown: the buyers have considered reactors to be unique forms of capacity. The pertinent finding is that total constructed or ordered nonbreeder capacity depends to a relatively minor extent on the charge made by the builder, as shown in $Q_n/Q = f(S, \mathrm{PNK/PFK}, \mathrm{PNK/PFF})$.[41]

The second conclusion from the history of nonbreeder reactors is that the firms carrying out the research and development effort are the ones that construct the commercial reactor plants. One reactor development program gestates one producer of commercial reactors;[42] following after the two nonbreeder programs which resulted in successful demonstration reactors, two companies — General Electric and Westinghouse — have had control of the supplies of thermal reactor capacity.[43] An extrapolation of history would require that every breeder program that results in a commercially acceptable reactor also results in a single manufacturer of that reactor system.

Then each of the new reactor types will be provided by a producer who is going to be a "new entrant," in some sense, into established and somewhat unique markets for reactors. The prices set by the entering producer are going to depend greatly on who he is. If he is one of the two controlling manufacturers of nonbreeder reactor plants, then it can be expected that the new type will be offered at prices "in line" with existing prices. The new price per kilowatt will deviate from the old only insofar as marginal costs dictate a change; in effect, if the price-cost margin PNK − MC** depends on the number of sellers and on demand elasticity $(\mathrm{PNK} - \mathrm{MC}^{**})/\mathrm{PNK} = -1/ne$, where n is the number of controlling sellers and e is the price elasticity of capacity demand, then the pricing decision is made with respect to the two established firms and the established demand elasticity. But if the entering

[41] Cf. the computed price elasticity in Q_n/Q with that from assuming perfect substitutability between nuclear and fossil plants at equal total generating costs, as in United States Atomic Energy Commission, *Forecast of Growth of Nuclear Power* (Washington, D.C.: U.S. Government Printing Office, December, 1967), where the implied price elasticity is numerically five times greater.

[42] Cf. Appendix A, where this finding is elaborated on in some detail.

[43] Cf. Appendix A; to have control of total reactor supply does not necessitate making all of the sales — as the many current studies of sales concentration and price-cost margins show. Cf. L. Preston and N. Collins, *Concentration and Price-Cost Margins* (Berkeley, Calif.: The University of California Press, 1968).

firm is not one of the two thermal reactor producers, then his pricing decision will depend on his forecast of the degree of uniqueness of the new type over the old nonbreeder reactors, and on the existence of three firms. The forecast here is that any new manufacturer, such as Atomics International or Gulf General Atomics, cannot expect any additional degree of uniqueness, and consequently has to treat demand after the advent of a single breeder system as if it will be shared among three firms of equal size. The two established producers dictate price-cost margin PNK − MC** $= f(n, e)$ by requiring that "n" be equal to or exceed three, and that demand elasticity "e" be for their joint demands for reactors. This is to expect that price formation on reactor systems can be described by two equations when there are breeders, $Q \cdot (Q_n/Q)$ for demand and (PNK − MC**) $= f(n, e)$ for price setting — where "n" depends on the degree of "newness" of the producer of breeder reactors.[44]

The Supplies and Demands of Fuel for Breeder Reactors

Since the core of the fast breeder is mostly plutonium oxide or carbide, with the coolant and "breeding" uranium of secondary importance, the markets for plutonium are the important fuel markets. There can scarcely be said to be such markets at the present time. The supply conditions are arbitrary, and with demands unknown, extrapolation of present conditions is impossible.

The supply, in effect, consists of the stocks of plutonium accumulated by the Atomic Energy Commission from the refining of "spent" cores from thermal reactors. These accumulated tonnages will depend on the demands for electricity in the next 15 years, and on the share of this generating capacity in nuclear-fueled systems, since these systems will

[44] Here MC** = C^{**}/kW, the marginal costs shown above. The argument is central to the research literature in the industrial economics of entry of new firms. The conceptual foundation of the approach is that of J. S. Bain, in *Barriers to New Competition* (Cambridge, Mass.: Harvard University Press, 1958), as formalized by F. Modigliani in "New Developments on the Oligopoly Front," *Journal of Political Economy*, Vol. LXVI (1958), pp. 215–232.

The approach taken here is to forecast the entry effects on reactor price-cost margins with the Cournot model, rather than the Bain model. The Cournot model is based on quite different behavioral assumptions (cf. E. H. Chamberlin, *The Theory of Monopolistic Competition* (Cambridge, Mass.: Harvard University Press, 1958), Appendix A), but it generates *approximately* the same predictions. It is preferred because it offers the opportunity to use the formal apparatus (PNK − MC** = − PNK/ne) in the computer simulation, where there is no such formal apparatus with Bain. The loss incurred from doing so is the approximation of the Bain forecast, as shown by F. M. Fisher, "Cournot and the Bain–Sylos Analysis," *Journal of Political Economy*, Vol. LXVII (1959), pp. 410–413. But the quality of the information processed here to form a most preliminary forecast of price effects is not sufficiently fine to detect the difference.

produce plutonium as a by-product of refueling and they will refuel only if they have been generating electricity. The construction of additional thermal reactor capacity should lead to more than 100,000 MW_e in place by the mid-1980's; each megawatt, given present patterns of in-core fission and energy conversion, should produce from 0.15 to 0.4 kg of plutonium per year so that from 15 to 40 T of plutonium should be available each year for use either in defense systems or in new breeder reactors. By the mid-1980's, there should be more than 150 tons from accumulation over the previous 15 years.

But stocks are not "supplies at various levels of prices." The Atomic Energy Commission — or one to three private companies, if the three diffusion plants are sold to private industry — can decide how much to charge for the stocks and how much to commit. They will probably not set prices so as to bring about a stock outage and excess demand, so that breeder owners are without fuel cores to operate the reactors. The assumption here is that price will be sufficiently high to prevent stock outage, where the plutonium stock $ZP = \delta_1 + \delta_2 Q_n$ with δ_1 the 1985 stocks and δ_2 the forecast lifetime stocks from thermal capacity Q_n added after 1985.[45]

The demands for plutonium are set by the marginal productivity of this fuel in sales revenues of reactors, at least in the world in which producers and buyers both can choose the efficiency of fuel consumption as well as the power rating of the reactor. As a first approximation, this is not assumed to be the case; rather, demand is equal to the stocks of plutonium per megawatt required to run a reactor similar to those in the design studies times the number of megawatts of rating.

In summary, then, the stocks of plutonium held by the refiners of spent fuel cores are assumed to be available to meet the "outside" fuel requirements of breeder reactor owners. The "inside" requirements come from bred fuel in the additional liquid metal or gas plants. The stocks are expected to be balanced against demands. The mechanism is price — the price of fuel setting the level of total breeder reactor capacity, which utilizes neither more nor less than these stocks.

The Forecast Benefits from Liquid Metal Fast Breeder Reactors

The development and sale of liquid metal breeder reactors would have extensive effects on the sales of both fossil-fueled generators and nonbreeder reactors. These reactors, ready for commercial adoption after 1985, would most probably be sold at higher capital prices but

[45] δ_2 is the rate of annual production per MW_e (from 0.2 to 0.4) capitalized over 30 years. Very roughly, $ZP = 150{,}000 + 1.88 Q_n$ kg, or $ZP^* = 150{,}000 + 3.76 Q_n$ in the more optimistic case.

with the promise of lower fuel cycle costs than the orthodox light-water systems — and the difference in capital prices is not forecast to be great enough to prevent the LMFBR from capturing almost all of the demands for reactors. To show this and the resulting consumers' and producers' benefits, consider the case in which there is "low" demand for reactor capacity of all types and in which the producer of the LMFBR is unrelated to the two large producers of nonbreeder reactors.

The demand function for nonbreeders in this case has been described, and the function for breeders can be assumed to be the same — as set by 2 per cent annual reductions in fossil fuel costs, and 1 per cent reductions in fossil and nuclear capital costs under those now in effect. The price equation for a new entrant is set by marginal costs, the elasticity of demand with respect to price, and the presence of three firms with some effect on the general level of reactor prices. The markets for fuel are cleared by a predetermined plutonium price — here, between \$5 and \$10 per gram — which sets the marginal costs close to \$325 per kilowatt for reactors in the 100–250 MW size class, \$158 per kilowatt for the 300–800 MW class, and \$116 per kilowatt for the 1,000 MW reactors. With price elasticity of -1.03 from $Q(Q_n/Q)$ and three price-setters, the prices are expected to be 50 per cent greater than these costs.

Under these circumstances, the price structure greatly favors sales of capacity in the LMFBR. There are three effects forecast from the four-equation simulation of the 1985–2004 period. First, the sales of plutonium at the present price not only clears the market of accumulations of this fuel in the 1970–1985 period, but also stabilizes the fuel cycle costs of the LMFBR between 0.5 and 0.8 mill/kWh. Second, the lower fuel cycle costs — by at least 0.5 mill/kWh — with the higher capital costs should shift almost all of the nuclear shares to breeders. Third, there is considerable additional capacity in reactors that would go into oil-, gas-, and coal-fired boilers in the world with only nonbreeder reactors.

These forecasts are shown in Table 3.6, and can be compared with those for nonbreeder reactors in Table 3.2. The breeders capture a good measure of the demands for reactors, and of the demands for nonreactors as well. Total dedications to the new reactor systems take 140 per cent of those of nonbreeders over the period. Total sales of reactor capacity increase by 280,000 MW, while fuel cycle costs are stabilized at 0.5 to 0.8 mill/kWh.

The full measure of benefits is not shown by the difference between capacity additions Q_0 for the light-water reactor systems and Q_1* for liquid metal and light-water systems together, but by some part of the

Table 3.6 Demands for Liquid Metal Breeder Reactors and Fuel Cycle Costs

Period	Additional Breeder Capacity (10^3 MW)	Additional Nonbreeder Capacity (10^3 MW)	Breeder Capital Prices* ($/kW)	Breeder Fuel Cycle Costs* (mills/kWh)
1985–1989	86	12	160	0.7
1990–1994	122	16	160	0.7
1995–1999	173	22	160	0.7
2000–2004	243	30	160	0.7

* For the 1,000 MW reactor only, assuming the price of plutonium is $5 per gram.

SOURCE: Estimates from the simulation of the four-equation model of fuel costs, reactor prices, and demands for capacity which is outlined in the text.

area between the two demand curves for these two alternatives. The area under the nonbreeder demand curve can be measured, assuming that such demand is

$$Q = f(P, Y/N, N) \quad \text{times} \quad Q_n/Q = f(\text{PNF, PNK, PFF, PFK})$$

for the relevant prices, and that Q_0 equals the additions to nonbreeder capacity in each power region while P_0 equals the light water reactor average price for the assumed size distribution of plants. The same area can be measured for the additional capacity accounted for by liquid metal breeders by assuming the same demand curves and the appropriate breeder prices, with Q_1^* as shown in Table 3.6 but for each power region and year and P_1^* the average liquid metal-cooled breeder prices of each size class. The net benefits are the differences between the two areas minus the marginal costs of providing the new breeder capacity.

Not all of the area of benefits can be measured or assessed reasonably at this time. Certainly the differences in demands above the going range of nonbreeder and forecast breeder prices cannot be measured, for statistical and economic reasons. This region of the demand equation is out of the range of the observations used to fit the least-squares function, so that forecasting is subject to wide error, particularly when using a logarithmic expression; the difference in demands may very well not be a measure of economic benefits, since it indicates only how much more the generating companies would have paid for less capital in order to generate more regulated profits. These amounts are hypothetical, since they assume passive allowance by the regulatory commission of rate base magnification, and they cannot be considered benefits for the final consumer of electricity in any case. They are not called benefits here (even though they are results specific to innovation with normal consumer demands).

The rest of the area can be measured, at least as an initial attempt to assess breeder reactor benefits. The measurement is in two parts. The first is the price-marginal cost difference for the additional reactor demand ($Q_1^* - Q_0$) for an approximation to producers' benefits. The second is the integral of the demand function between Q_1^* and Q_0 above price P_1^* as an approximation to consumers' benefits.[46]

The total of these benefits in the 1980's and the 1990's is very likely to be at least $79 billion. As shown in Table 3.7, the modest additions

Table 3.7 The Forecast Benefits from Liquid Metal Breeder Reactors

Period	Benefits, assuming "low" demands for both breeder and nonbreeder reactors (billions of dollars) Producers' Benefits	Consumers' Benefits
1985–1989	2.5	10.0
1990–1994	4.2	19.2
1995–1999	6.5	32.1
2000–2004	9.7	51.2

SOURCE: Areas under the relevant demand curves labeled CSBB and CSBC as in Appendix F, or as described in the text.

to demand in the early years limit the gains to close to $12 billion in the first five-year period after introduction of the new technology. But in the period right before the turn of the century, the gains from the new breeder system exceed $39 billion. The additions to population and per capita income forecast for the first five years of the twenty-first century add greatly to demand in that period; as a result, the capacity in nuclear reactors is much greater than in earlier years and adoption of the breeder for that capacity promises larger benefits. The consumers' benefits should exceed $9 billion, and the producers' $51 billion, in that five-year period alone.

[46] These areas are designated CSBB and CSBC in the simulation program and in Appendix F. Area CSBA is not calculated for the statistical and economic reasons given above.

The Forecast Benefits from Gas-Cooled Fast Breeder Reactors

The second possibility is that the gas-cooled breeder reactor will be developed instead of the liquid metal breeder reactor. The inauguration of a program for developing gas cooling could well begin within the next three years. The completion of the program cannot be promised to take place at the same time as that of the alternative program, however. The best estimate at this time is that the demonstration reactor based on a gas system would show commercial possibilities during the late 1980's. Then the economic benefits from installing gas breeders would be realized in the 1990's and the early years of the twenty-first century.

The gas breeder, as in the case of the LMFBR, might be the third independent source of supply of reactor technology, with the other two sources being accounted for by the two major light water reactor manufacturers. Setting prices for this breeder would be a matter of price elasticities of demand $Q(Q_n/Q)$ and the presence of two other sources of reactor supply (as well as the marginal costs of gas-cooled plants of the desired sizes).[47] The pricing policy of the manufacturer of the gas-cooled plants, based on the existing supplies of the established manufacturers, is expected to reduce prices for all sizes of plants below the general level of those for the LMFBR case: marginal costs are lower, because of lower equipment costs and lower fuel costs. This last reduction — in the prices of plutonium and the consequent costs of fuel consumed in mills per kilowatt hour — is especially striking; it follows from the supplier of plutonium having to sell at between $5 and $10 per pound to clear the market of not only the accumulated output of thermal reactors, but the net additional output of the first breeders in the early 1990's. The prices of capital to the reactor purchaser and accompanying fuel cycle costs are shown in Table 3.8. They imply sales of 730,000 MW in the short time period between 1990 and 2004.

The effects of these prices on the net benefits from breeder reactors are shown in Table 3.9. There are no measurable gains in the 1985–1989 period because of the late introduction of this type of reactor, but there are approximately $95 billion in buyers' and sellers' benefits in the 1990's and $90 billion in the first five years of the new century from this set of forecast prices and costs. The gains from the lower marginal costs and prices are greater in any one time period than from

[47] The fuel cycle costs on the demand side consistent with those shown above in the marginal production costs are somewhat less than 0.5 mills/kWh. This estimate is used as PNF; the fossil plant prices PFF and PFK are the same as used to estimate the gains from the liquid metal breeder reactor.

Table 3.8 Demands for Gas Fast Breeder Reactors and Fuel Cycle Costs

Period	Additional Breeder Capacity (10^3 MW)	Additional Nonbreeder Capacity (10^3 MW)	Breeder Capital Prices* ($/kW)	Breeder Fuel Cycle Costs* (mills/kWh)
1985–1989	—	†	—	—
1990–1994	166	18	147	0.4
1995–1999	234	24	147	0.5
2000–2004	330	33	147	0.5

* For the 1,000 MW reactor only.
† Equal to total thermal capacity in a world without breeders, as in Table 3.2.
SOURCE: As in Table 3.6 and preceding tables.

the alternative fast breeder project in that time period. But the gas breeder will take five years longer to develop, so that it loses all the benefits accruing to the LMFBR in the 1985–1989 period.

An Initial Comparison of Benefits

The two independent and separate types of breeders can be compared in terms of the benefits they provide to the producers and buyers of generating equipment. A first attempt to do so could be based on Tables 3.7 and 3.9; they show the resulting producers' and consumers'

Table 3.9 The Forecast Benefits from Gas Fast Breeder Reactors

Period	Benefits, assuming "low" demands for both breeder and for nonbreeder reactors (billions of dollars)	
	Producers' Benefits	Consumers' Benefits
1985–1989	*	*
1990–1994	5.9	31.2
1995–1999	8.7	49.7
2000–2004	12.7	76.9

* Forecast as negligible since the demonstration plant program will not be complete.

SOURCE: As in Table 3.7.

surpluses for only one of a number of cost and demand conditions (the others will be analyzed in the next chapter), but general differences are apparent. The liquid metal breeder reactor system has higher prices and fuel costs, and thus generates less buyers' demands and lower benefits. This system is ill-adapted to providing the small plants in the size distribution at prices similar to those of light-water reactors and fossil fuel boilers; then the share of breeders in the smaller size classes is especially small. The second type of breeder reactor comes much closer to stabilizing fuel cycle costs at levels close to 0.5 mill/kWh, in a gas-cooled plant with capital prices less than $150 per kilowatt at full size. At one-half and one-third size, the gas reactor does not have the price disadvantages of the liquid metal reactor because of the simplicity and divisibility of its system, so that there is at least a token gas breeder share of capacity in the small size classes. The gas benefits are greater in 1990–1994 and subsequent periods as a result. But there are no gas benefits earlier, while there are substantial benefits from a liquid metal breeder, given the longer gestation period of the gas-cooled technology. On the whole, the question of relative benefits will have to be answered in terms of the net present values of these two types of reactors. These net values are assessed in finding the optimal economic strategy in the next chapter.

4

The Economic Fast Breeder Development Program

The research costs for developing various types of breeder reactors and the subsequent economic benefits from installing commercial copies of these reactors set the economic strategy for the next few years. When the present value of forecast costs exceeds that of total benefits from successful completion of the project, it is clear that the economics of decision-making requires the termination of this research. The value of the resources used on the project, as represented by total payments or expenditures on the research necessary to draw resources away from the other uses, is greater than the buyers' and sellers' benefits above that which they can obtain from the other uses.[1] In other cases, when total expenditures do not exceed the gains from research, then the choice among projects might be made on the basis of relative net present benefits. The project promising the greatest gains for given dollars of research expenditure — that is, the highest net present value — is the best project from an economic point of view.

This economic strategy can be laid out for the first round of decisions on the number and types of fast reactor projects. But comparing separate projects, and combinations of projects, is a matter of comparing the distributions of net present values generated by the research costs and gains in each case.

The initial survey of the costs of research in Chapter 2 arrived at the conclusion that three levels of expenditures are equally probable. The presence of research breakthroughs could result in the occurrence of

[1] This would seem a necessary, but not sufficient, condition for considering a project worthy of support. The gains as measured here are social benefits if there are no further costs imposed on others by this activity, if the resulting distribution of benefits is optimal, and if the required rates of production here do not add to misallocation of resources in this industry and elsewhere. These conditions might hold for the forecast capacities shown in the last chapter — at least there is a finite probability that they hold. But they cannot hold if the gains are less than the research costs.

"low" research costs at one stage without the same fortunate circumstances occurring at other stages; "target," or even "high," costs could be experienced as a result of planned breakthroughs or unplanned setbacks at that stage. With four stages — fuel development, core development, components, and systems development — and three possible results at each stage, there are 81 probable levels of total research expense. The frequency distribution of costs of research can be expected to show this range of results for each project.

The benefits forecast for each type of commercial fast reactor vary over a fairly wide range, as well. The demands for the liquid metal fast reactor, for example, are expected to be "low" if there is reasonable technical progress in nonbreeder and fossil-fueled systems which result in price reductions for these systems; these losses to other technologies imply reduced reactor benefits as compared to those from the "higher" level of demands. The probability of the lower demands is believed to be roughly equal to that of the higher demands in the cases of the gas-cooled and sodium-cooled breeder reactors. The same conditions apply to nonbreeder versus fossil plants, so that the breeder and nonbreeder demand combinations result in two levels of probable benefits from breeder over nonbreeder reactors. That is, "high" demands for breeders over and above "high" demands for nonbreeders result in one forecast level of benefits, while "low" demands for both result in the other forecast. This is the case for both breeder types.

There is an additional source of variation in benefits in the case of the liquid metal breeder. The costs of operating this breeder vary more with the price of plutonium because of higher inventory requirements and lower breeding rates; considering the range of possible Pu prices within $3 per gram of the forecast market clearing price — the costs, prices, and consequent benefits take on a wider range of values. Here four values of benefits are estimated rather than two, with two each for high demand ("low" and "high" fuel costs) and two each for low demand.

Both research costs and benefits have to be discounted by the proper rate of interest to obtain net present value. But what is the proper rate of interest? The most general and plausible argument is that this rate r^* is the opportunity cost of capital in this large-scale project or set of projects, where the investments neglected in favor of the development of the breeder would have earned r^* per cent. That is, "for most public investment, use the private returns foregone to release funds for public investments, as the appropriate rate."[2] This is justified on the

[2] Cf. "Remarks" of J. S. Hoffman, Assistant Director of the Bureau of the Budget, before the Subcommittee on Economy in Government of the Joint Economic Committee (July 30, 1968), p. 26.

basis that a Federally sponsored research program should not require private citizens "to give up a portion of their incomes in the form of higher taxes to support public undertakings which are of less social value than the uses to which their funds would otherwise be put."[3] The actual rate cannot be less than 10 per cent per annum, and might well not be greater than 15 per cent; at least, "projects which displace specific private investments should be evaluated by the rate of return prevailing in the sector from which the specific investment is displaced" and the heavy electrical and nonelectrical equipment industries can easily be expected to earn in this range.[4] Both 10 and 15 per cent rates of discount will be used to find the range of probable net present values.

The costs C_t and benefits B_t in year t generate the net present value for a project equal to $V = \sum_{t^*}^{n} B_t(1 + r^*)^{-t} - \sum_{1}^{t^*} C_t(1 + r^*)^{-t}$ for the social rate of discount equal to $r = r^*$ on such research. Other levels of costs C_t^* and gains B_t^* generate different net present values. The values V associated with all levels of research costs and benefits make up a frequency distribution for a particular reactor project. There are distributions for the liquid metal and gas fast reactor projects from the range of forecast research costs and benefits for each.

The choice of the most economic projects in fast breeder reactor development comes to selecting those with the highest average and lowest variation in net present values. Questions of choice are almost unlimited; but there are a few that are most basic. If only one project is to be undertaken, then which type of reactor is most economic? Which company should undertake the research for the Atomic Energy Commission? If more than one project is conceivable, then how many, and how many competitive projects, should be carried out at one time? The first round of answers can be attempted on the basis of present knowledge and forecasts.

The Most Economic Single Reactor Project

The first application of economic strategy is to the choice of a long-term research project to build a demonstration plant of a single type of fast breeder reactor. The candidates discussed here are the LMFBR, the GCBR, or none. The one that has the most desirable forecast distribution of net present values is "economic."[5]

[3] Cf. Joint Economic Committee 99-416 O, *Economic Analysis of Public Investment Decisions* (Washington, D.C.: U.S. Government Printing Office, 1961), p. 10.

[4] *Ibid.*, p. 12.

[5] The choice is extremely limited, since it is constrained by the necessity to hold the number of projects to a single endeavor, not by the total amount of capital involved in research — since these amounts differ from project to project — or by

This "economic" project is shown by extensive computations of initial forecasts of costs and benefits, similar to the examples in the last two chapters. The costs of research for each project, evaluated at 81 probable levels of expenditure, set against each of two to four probable levels for dollar benefits from research, are the basis for 162 to 324 computed net present values for the project.[6] That is, a distribution of values of V is calculated for each reactor by setting each of these research costs against the benefits in present value terms. The different distributions are summarized in Table 4.1 over probability intervals (where the probabilities are the products of those for a particular level of research cost and for a particular level of gains).[7]

The two sets of distributions of returns are revealing. The LMFBR quite evidently offers the lowest expected net present values. The

Table 4.1 Forecast Net Present Values from Each Breeder Reactor Project

	Net Present Value V^* (billions of dollars)			
Probability of	GCBR		LMFBR	
$V < V^*$	10% Rate of Discount	15% Rate of Discount	10% Rate of Discount	15% Rate of Discount
0.01	12.8	3.5	9.3	2.3
0.10	13.0	3.7	9.5	2.5
0.20	13.2	3.8	9.7	2.6
0.35	13.4	3.9	9.8	2.8
0.50	16.3	5.0	9.9	3.1
0.65	16.6	5.2	10.0	3.2
0.80	16.9	5.3	10.2	3.3
0.90	17.0	5.4	10.2	3.4
0.99	17.2	5.6	10.5	3.6

SOURCE: Calculated as outlined in the text.

the time period required to carry out the research. But the rationale is that anything less than a single project accomplishes nothing toward reducing nuclear fuel utilization in the last years of the twentieth century. (That is, the projects of smaller size than full development of a fast breeder type must promise zero or negative economic returns.) The rationale for more than one is explained in the next section.

[6] The totals for each five-year period are assigned to separate years by a straight-line interpolation between them for both the case of high prices as shown in Table 3.6 and low prices in Table 3.7.

[7] The discussion in Chapter 2 is to the point that the probabilities of "low" and "high" research costs equal the probability of "design" research costs. The four levels of gains for the liquid metal fast breeder and the two for the gas fast breeder seem equally probable.

average or expected net value is $9.9 billion, while the range of probable values is $1.2 billion in this distribution at the 10 per cent rate of discount. This again indicates that the liquid metal technology is ill-adapted to meeting the demands for a size distribution of plants that has a large percentage of small plants.[8] But the distribution of net value from the LMFBR also shows that the range of possible research costs is relatively wide and the range of benefits is greatly widened by possible plutonium fuel price variation late in this century. The gas-cooled breeder reactor promises more gains from research than the liquid metal breeder. The average net value is forecast at $16.2 billion, while the range of values is $4.4 billion. The average is greater, even though the gas technology arrives on the scene some five years later. The range of results is greater, but none of the probable results are as low as all of the probable net values from the liquid metal breeder reactor.

The most economic project would seem to be the GCBR program. The commercial gas reactors of various sizes should meet the demands at lower prices and costs for all sizes of reactors and should account for more than half the plants in both medium and large size classes. The costs for developing commercial gas reactors are lower than for liquid metal breeder reactors. As a consequence, the distribution of net present values forecast for the gas-cooled breeder has consistently higher estimates than that for the other potential breeder reactor.

The Number of Reactor Manufacturers

The financing of a single breeder reactor project does not by itself determine the number of firms producing nuclear energy plants. There are two large-scale and two or more small-scale manufacturers of nonbreeder reactors who can be expected to continue to sell their equipment after the completion of a breeder development program. If they are financed to perform the new research, individually or collectively, then there will be no new companies; but if an attempt is made to underwrite the new expertise in other companies, then more firms will be added to the production side of the reactor market. Should such an attempt be made?

There is some basis for arguing that a third large manufacturer, rather than new reactors from one or two of the established nonbreeder reactor producers, would add substantially to the net present value of

[8] The LMFBR accounts for only 12 per cent of capacity in the smallest size class of plants over the period, but 40 per cent of capacity in the plants with more than 1,000 MW_e, according to simulation of "low" demand conditions.

the breeder program. The new firm would be faced with "breaking in" with new production technology, as would the established producers of new breeders; but the new firm would also have to add to the output of the existing producers, while the established firm can be expected to replace a major part of nonbreeder offerings with breeders. The price forecast, in the case of an independent third supplier, would have to center on $160 per kilowatt for new liquid metal fast breeder capacity at 1,000 MW, but $210 per kilowatt when from the established firm.[9] The rate of adoptions of liquid metal breeders would be from 32 to 33 per cent of the medium-sized plants, and 53 to 54 per cent of 1,000 MW_e plants, when purchased from the independent manufacturer, but only 20 to 24 per cent for the medium-size class, and 37 to 38 per cent for the larger class, from the established manufacturer (as shown by the simulation of the entire 1985–2004 period).

The higher prices and smaller quantities result in substantially less consumers' and producers' benefits. The total of benefits with two firms setting prices are $8.5 billion less than from three large firms in the first five years, $13.0 billion less in the second five years, and $19.1 billion less in the last five years of the 1985–1999 period. The benefits lost by the higher prices and reduced breeder share of total capacity demand come to $27.7 billion in the years between 2000 and 2004. The present value of these benefits is $6.1 billion at the 10 per cent rate of discount, which comes to 55.8 per cent of the total benefits from the LMFBR under "high" demand conditions.[10]

The only measure for ensuring that the greater benefits will be realized, in the case of the LMFBR, is to have the government-sponsored research done by one of the companies now carrying out small liquid metal experiments, but not now manufacturing thermal reactors. There may be substantial additional costs of research; but these cannot exceed $6 billion in net present value. The most likely program for the development of gas-cooled reactors is different. The largest and most expert research organization in gas technology does not now produce thermal reactors to a scale competitive with General Electric or Westinghouse. Entry by a "new" producer is the inevitable result of a successful gas

[9] Both forecasts are made from the price equation $PNK = MC[ne/(ne + 1)]$ where $n = 2$ for breeders from established reactor manufacturers and $n = 3$ when breeders come from an entirely new firm. These are alternative parameter values in the four-equation model, so that they result in different reactor prices, plutonium demands and prices, reactor capacity, and ultimately, consumers' surpluses. They are based on the assumption that, to enter, a new firm would have to reduce his (and other firms') prices to an extent in keeping with the Bain worst conditions as described above, while the established firm would enter with "best" conditions.

[10] The present value is $2.2 billion at a 15 per cent discount rate — 65.7 per cent of net present value from the LMFBR at that discount rate as shown in Table 4.1.

reactor program. The identity of the producer, then, is a second economic reason for priority for gas research: lower prices and greater benefits can be expected from merely carrying out the research in gas, since this adds to the number of large sources of supply of reactors.

The Economic Combination of Projects

The fast breeder reactor development program could be limited to a single set of research activities designed to produce one type of commercial demonstration reactor. But it is conceivable that history could be repeated by requiring that the Atomic Energy Commission underwrite initial research once again on a number of separate types of nuclear reactors. The LMFBR development program, underway since the early 1960's, could be continued at the same or a faster pace; the development of gas-cooled reactors could become a major new federal project by enlarging the private company programs now under way to a scale equal to that of the liquid metal research program. Other technologies not discussed here, or not yet beyond theoretical formulation, could provide the basis for building commercial demonstration reactors in the mid-1990's or later. All of these could be undertaken at once in the tradition of thermal reactor development: start as many projects as can provide plans, and continue them until the difference between forecast and realized research costs of one or two projects is significantly less than for the others; then eliminate the other research projects in favor of these one or two promising programs.

The questions of strategy raised by these possibilities are sharply delineated. The technologies of liquid metal and gas fast reactors show possibilities for rapid and safe development — at costs which vary within the ranges shown in Chapter 2. There is some indication that not only can these reactors be developed, but a "consolidated" large-scale testing and construction program of the two together promises more results than those for either one of the independent programs for a reactor type. Not only do forecast costs indicate advantages for "consolidated" research, but also the forecast benefits from commercial breeders could well be greater when there are "competitive" research projects. Then the choice would be between the single exclusive project with the greatest net value and a larger number of projects parceled out among commercial companies so as to maximize competition in the final market for reactors.

Choosing the number of reactor projects can be put in terms of economic results. The number of projects and their arrangement resulting in the most favorable forecast distribution of net present value is most likely two projects parceled out so as to ensure competition in the sale of the two new reactor types. This can be seen from the contrasts between net present values on two reactors as a consolidated project and those on the exclusive single reactor project.

Competitive Fast Breeder Development Projects

Two new types of fast breeder reactor could be added to the alternatives available to the electricity generating companies in the 1980's and 1990's. The first could come from that company building the experimental and demonstration LMFBR reactors. The second would come from a company working in gas technology which, when given results from fast flux testing of breeder cores in the general research program at Hanford, Washington, builds a gas-cooled fast reactor. The liquid metal breeder demonstration plant could be constructed successfully in the mid-1980's, and the gas-cooled plant in the late 1980's. Both could be built by separate and independent reactor manufacturers — companies other than the two large light-water reactor manufacturers — now carrying out research in these two technologies, and contracts could be let to give these companies a strong lead in further development. This would maximize the number of competitors.

The economic benefits from competitive development projects differ from those expected from single and exclusive projects. The costs of research on a reactor type are those shown in the "consolidated" two reactor program outlined in Chapter 2. But the benefits from commercial adoptions following research are different from those in the single project because reactor sales prices are different. Prices can follow any number of patterns when there are four large reactor manufacturers, each offering plant types over a range of sizes in each power region. Prices might be maintained as if there were still only two firms, given the loyalty of the third and fourth firms to the policies of the established manufacturers; in this case, the only gains from breeders are found in increased buyers' demands from lower fuel cycle costs. To the other extreme, the presence of four firms might break the pattern of loyalty and establish approximately perfect competition. Then prices would approach marginal costs and the buyers' benefits would be a maximum (for those marginal costs and for the costs of research resulting from this scheme of separate projects). Neither extreme seems as likely as the continuation of the current history of reactor pricing; but with the producers of breeders having to offer price concessions to enter the market with two new technologies, the number of new technologies is important. It is expected to be necessary for the new companies to accept lower profit margins when there are twice as many entrants, and the initial forecast is that the price per kilowatt PNK $=$ MC$[ne/(ne+1)]$ will be determined by $n = 4$ rather than $n = 2$ or $n = 3$ for the total number of reactor producers.

Consider the liquid metal fast breeder program as chronologically "first," appearing to have the capability to construct a working demonstration reactor in the early 1980's. The gas breeder project might follow the liquid sodium project with a demonstration plant showing good commercial possibilities in the late 1980's, and with commercial copies available at prices reduced to make sales in the presence of two established light-water reactors and the newly established liquid metal-cooled breeder reactor. The benefits from doing so would be equal to the *additional* buyers' surplus generated by lower fuel cycle costs and lower capital prices. These gains, as the integral of the demand function $Q(Q_n/Q) = f(\text{PNF*}, \text{PNK*})$ for lower fuel and capital prices PNF*, PNK* minus the integral of the same function for prices for the thermal or nonbreeder reactor PNF, PNK, are as follows:

Time Period	Benefits (billions of dollars) "Low" Demands	"High" Demands
1985–1989	12.5	23.7
1990–1994	41.5	53.9
1995–1999	64.7	87.7
2000–2004	98.7	137.1

The benefits far exceed those associated with this liquid sodium breeder as a single project. It would seem that reduced reactor prices and fuel cycle costs together are more than compensated for by increased demands. There are benefits, as well, in cost reductions for producers, given the lower costs of building gas, rather than liquid metal breeders. Balancing "high" and "low" benefits against the total or "consolidated" costs of research on two reactors results in the distribution of net present values shown in Table 4.2. These forecast net values from two projects come to $18.3 billion on average, with a range of $5.0 billion. The average is almost twice as great as when the LMFBR is the only breeder reactor project, and the range includes none of the estimated values of the LMFBR as the single project.

What are the results from two "competitive" fast reactor programs? The benefits from two projects are expected to derive from the increased nuclear shares generated by lower prices following from adding two

Table 4.2 The Forecast Net Present Values of Two Competitive Breeder Reactor Projects

Probability of $V < V^*$	Net Present Value V^* (billions of dollars) 10% Rate of Discount	15% Rate of Discount
0.01	15.3	4.6
0.10	15.6	4.8
0.20	15.8	5.0
0.35	16.0	5.1
0.50	19.0	5.6
0.65	19.7	6.1
0.80	19.9	6.3
0.90	20.0	6.4
0.99	20.3	6.6

SOURCE: Calculated as outlined in the text.

more reactor manufacturers. The producer of liquid metal reactors would almost certainly be the firm carrying out the last stages of research in sodium technology. This should be a company separate and independent from the two light-water manufacturers. The same would be true of the producer of gas fast reactors — this company would specialize in this technology in order to make the number of competitors as large as possible. Increasing the number of alternatives would result in "passing through" the lowest fuel cycle costs — less than $\frac{1}{2}$ mill/kWh under the low-cost conditions associated with generating large supplies of plutonium in the gas reactor for use in liquid metal reactors. The increase in alternatives would give the buyer the chance to select the fast reactor type best able to provide the plant size needed to make up his required size distribution of plants. Capital prices on the small- or medium-scale gas plants would be expected to be more attractive, and capital prices on the large plants in both technologies are expected to be equally attractive. The prices on the middle-sized and large-sized gas plants and the largest liquid sodium plants would be "competitive," in the sense that they would more closely approximate producers' marginal costs.

The policy to foster duplicative research is not extreme. Examination of the policy indicates that buyers' gains increase from moving in this direction. Gains accrue from setting out more programs with higher total research expenditures, which result in more independent producers of distinct types of breeders — because the reductions expected to take place in margins between reactor prices and costs result in lower prices even with higher research costs.[11]

Decision for the Next Decade, Given the Expected Benefits from Research

The Atomic Energy Commission has assigned the highest priority to "finding and executing the research and development program required for mastering the technically difficult and challenging sodium-cooled fast breeder reactor concept in a timely manner."[12] The application of economic strategy suggests a reappraisal of this priority and a considerable extension of the program beyond its scope.

The strategy calls for decisions based both on costs and on the benefits from developing new reactors. The costs of research of the liquid metal program are somewhat greater under most conditions than those expected for the gas-cooled breeder program. If costs are dominant, so that any single program is limited to the expenditure level forecast for the LMFBR, then the liquid metal project offers less. If only one project can be chosen without a prior ceiling on expenditures, then the LMFBR offers less in net benefits. The probable net present values from spending the required amount for a single gas-cooled project are greater than for the liquid metal project — at all levels of probability. Development of wholly new gas-cooled fast breeder reactors promises more than the liquid metal project when promise is measured in terms of producers' and consumers' surpluses implicit in the sales of electricity generating equipment.

Where the goal is to achieve the highest net present value, then more than a single reactor project is of primary importance. The economic strategy requires priority for doing two projects "in parallel" so as to add competitors. The frequency distribution of net present values

[11] A remaining question is the timing of the research. In the tradition of large-scale dam or other reclamation projects, it may be economically efficient to announce the advent of a large-scale breeder research program at the present time, but to postpone financing this research for some years. There is no evidence of economic benefits from doing so. Any delay would have to be a five- to seven-year delay, with a three-year restart period, because present postponement would remove the potential independent breeder producers from this type of research (since the gains are in good part those from competition, re-entry time requirement is critical). The forecast net present values of benefits minus costs for the consolidated two reactor programs are $7 billion to $8 billion less at the 10 per cent rate of discount, and $4 billion less at the 15 per cent rate for all levels of probability, from a total delay of 10 years. This loss is calculated a number of ways — either by eliminating 1985–1994 benefits and postponing 1970–1979 costs, or by moving the whole program forward 10 years — and it would seem sufficient to discount any suggestion of postponement.

[12] Milton Shaw, "Fast Breeder Programs in the United States," *London Conference on Fast Breeder Reactors* (17–19 May, 1966: British Nuclear Energy Society), p. 1.

forecast for two independent fast reactor projects has strong advantages over that for the LMFBR as a single exclusive project. If a case can be made for one program and for the LMFBR as the Atomic Energy Commission program, then two independent programs present a better case. Most important, a case can be made for more projects, when the number of competitors is increased as a result. The benefits for buyers of generating equipment from more independent reactor manufacturers are greater than from a thermal reactor manufacturer offering some one type of fast breeder in appropriate sizes.

Economic strategy calls for favoring research on a much larger scale than is now being carried on. Programs should be underway to develop LMFBR and gas fast breeders of all size classes. Other technologies — such as steam-cooled breeding — should be continually scrutinized for evidence of promise equal to that of sodium and gas technologies. Strategy points to financing large-scale research and experimentation facilities to carry out the work on the best known breeder types. The two reactors, offered for overlapping size classes of plants, promise marginal net present values twice as great as those from the sodium-cooled project now assigned "the highest priority." For these gains to be realized, it is necessary to expand the size, the number of reactor types, and the number of firms in the reactor research program. For economic reasons there should be a return to the policies of duplicating projects and programs in different companies and technologies.

Appendix A

Strategies for Nuclear Reactor Development in the 1950's and 1960's

Programs for developing steam generators using nuclear energy were started in the early 1950's — perhaps much earlier, on an informal basis — for quite obvious reasons. The goal of the early programs was clearly to develop a system from capital equipment, nuclear fuel, and labor which utilized energy from nuclear fission to produce high-temperature steam. The principles of fission energy had been known for some time: by striking the nucleus of a heavy atom with a neutron, so as to split this nucleus into two or more parts, a reduction in mass takes place which has to be accompanied by the release of energy in accordance with the differential of $E = Mc^2$. But it was not practical to put this energy to use because fission could not be sustained, nor was equipment available for capturing a large portion of the resulting energy. During the late 1940's great progress was made on solving these problems, and since then the challenge has been to capture energy at costs less than those associated with that from boilers using fossil fuels.

Theoretical and Applied Research

For continuous energy release, fission must be self-sustaining — neutrons must be released by N fissioning atoms which collide with N atoms that are also fissionable, or else the complicated and unwieldly task of continually inserting neutrons into the fissioning material has to be undertaken. Not every neutron released in fission strikes another atom; some escape from the material by a chance lack of collision while others form isotopes of no importance or value. Then to ascertain that one released neutron collides with one more fissionable atom, a certain-sized volume of nuclear fuel is required. Few neutrons can escape, so that capture of neutrons by fissioning atoms is enhanced, if the fuel mass has a small surface area and a large internal fissioning area. To provide the best fissioning environment, this fuel mass or "core" should be as large as the state of the art will allow, because the escape surface is then the smallest possible in relation to the volume of fissioning material.

The fissionable atoms undergoing the greatest release of mass and energy are thorium Th^{232}, plutonium Pu^{239}, and the uraniums U^{233}, U^{235}, and U^{238}. The first and the last materials have a significant probability of fission only if neutrons collide with their atoms at high speed (with kinetic energy of more than one million electron volts) while the others fission with neutrons at all speeds but most probably with slow neutrons ($\frac{1}{10}$ of an electron volt). Then there is a choice of environments; the reaction can occur in a compact core of Th^{232} or U^{238} if fast neutrons cause fission, or in a diffuse core of U^{233}, U^{235} with a moderating material to slow down neutrons. The first is a *fast reactor core*, the second a *thermal reactor core*.

The choice of reactor core size and type depends on the state of research in applied physics. In the case of a design for a thermal reactor, the theory specifying core size required to sustain fission may exist, but in that of a fast reactor this work may not be done or may remain to be tested by laboratory experiments. As important as core size, is the specification of that core content which allows heat production without losing control of the reactor. Surges of temperature and pressure, even if short-lived, might in some instances accelerate fission so as to render any outside limits ineffective; theoretical analysis of these instances must be complete enough to make it possible to avoid them.

The choice of a reactor is broadened by the availability of a large number of potential transportation media for carrying heat energy from fuel core to steam generator. Each medium operates best under particular core temperature and pressure conditions, so that there is direct feedback from heat exchanger components to reactor internals. The routing of heat energy generally follows that shown in Figure A.1: the heat-absorbing liquid or gas is pumped up through the core and then out of the reactor to a heat exchanger; here steam receives the heat, is moved to the turbine and subsequently to the condenser. The coolant in the primary routing can be a liquid metal such as sodium, or a gas such as helium, or even steam itself.

The choice here is a matter of the technology of materials. For the reactor with moderating material a_1 and coolant material b_1, the least-cost combination of capital components, fuel, and labor dictates the use of metals able to operate with b_1 at 500°F and 1,000 psig. But another reactor type with a_2, b_2 operates best — with lowest costs for that type — with b_2 at 1,000°F and 100 psig. The reactor types proposed in the late 1940's could not have been built and operated for long

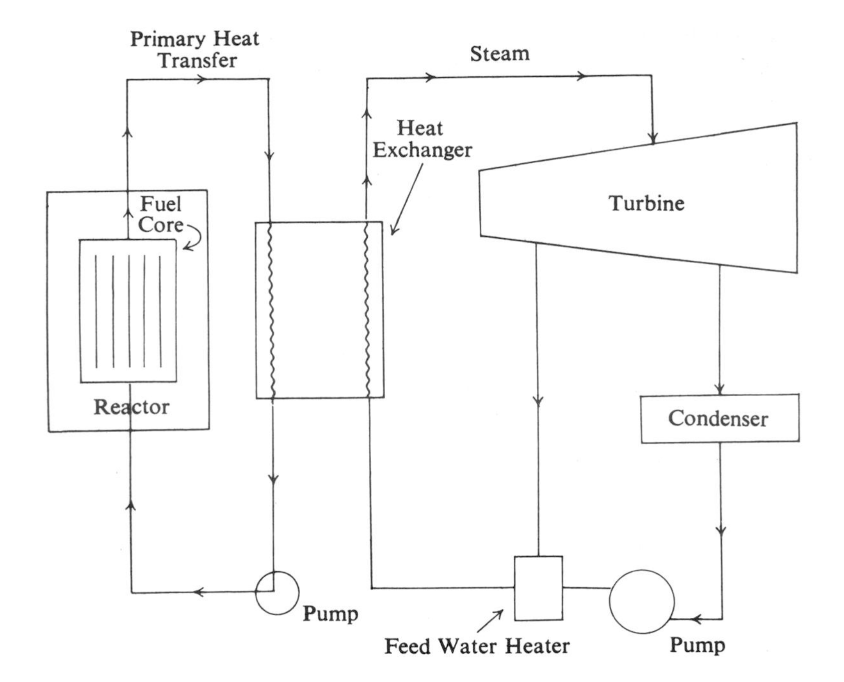

FIGURE A.1 Simplified component layout for a nuclear reactor

periods because there were not metals available that could withstand the stress and fluid corrosion of the first coolant at the first temperature and pressure, nor the corrosion and radiation damage from the second coolant under the second, higher temperature but lower pressure.

Research progress almost always called for the discovery and development of new metallic substances for reactor systems. These required experimentation with available but unproven substances, or new combinations of elements, under the specified conditions of temperature and pressure in a test facility built for a particular coolant. For sodium in a fast reactor core at 600°F and 100 psig, a test facility (termed Experimental Breeder Reactor I, or EBR-I) was constructed in 1951 at the National Reactor Testing Station in Idaho; for water as moderator and coolant at 417°F and 300 psig in a thermal reactor, a separate facility was constructed in 1954 (the BORAX III). Experimentation has more often than not been a matter of trying materials termed "likely" as a result of past observations of alternatives under somewhat similar conditions. There has been little theory to guide the experimenter: "The properties of materials cannot be predicted from known natural principles, as engineering performance can be. This is the heart of the present materials problem for nuclear reactors . . . experimental study is the only way to get the necessary answers."[1]

When solutions have been found both to the theoretical problems in neutron capture and to experimental problems in prolonging the life-time of materials, then reactor systems have been put together. The first attempt to do so has usually been a *reactor experiment* which shows that the interactions of test components is not unstable. The plant may or may not have been more costly to operate than alternative systems; almost invariably costs are not mentioned in the single-minded pursuit of continuous operating performance. But then *reactor prototypes* have been built on the basis of experiments, and operated to continue the earlier work in the direction of reducing component costs and improving component performance. When these plants showed performance of components with costs similar to those for the same components in other reactor types considered to be at an "advanced stage," then *demonstration reactors* have been constructed. Success at this level showed to the final users of energy-producing systems that this reactor type could be chosen on the next round of construction of new commercial facilities.

Strategy in the Early 1950's

Collection of information on uranium reactions obviously began in the United States in the atomic weapons development program during World War II; even though most of it was not pertinent to reactor development, enough was related to the applied physics of the fuel core that reactor research could begin immediately in 1945. The weapons program itself was transformed into a Federal reactor development program: the national laboratories at Los Alamos and Oak Ridge took the initial steps toward reactors in the war research facilities by constructing "mock ups" there of self-sustaining fuel cores with heat transfer routings to steam generators. The bomb program was dormant, personnel were interested, and funds were available to keep the personnel in the facilities from drifting away. Both the resident scientists interested in continuing the relevant basic physics research and the personnel in laboratories and workshops for components development gave the Federal laboratories a strong lead in building the first systems.

There was nothing that could be called a "plan" or "strategy" directing these early steps in experimentation. Projects of interest to the

[1] Dr. Russell Dayton, Battelle Memorial Institute, in testimony before the Joint Committee on Atomic Energy, 87th Congress, in *Development, Growth, and State of the Atomic Energy Industry* (Washington, D.C.: U.S. Government Printing Office, 1961), p. 88.

installed research groups were tried. Overall direction began with the organization of the Division of Reactor Development in the Atomic Energy Commission in February, 1949 and the first plan — consisting of four major reactor projects — was announced soon after. These were still based almost entirely on the interests and expertise of established research groups in national laboratories — two of the projects were continuations, the other two may have been on the basis of "demand conditions" separate from growing research interest.

The first two projects were supply-determined in the sense that previous laboratory research programs had produced test results in physics and working components which put them ahead in developing an operating prototype. The experimental breeder project had a lead at the initial stage of putting a system together, because a Los Alamos experimental fast reactor termed "Clementine" constructed on breeding principles had been working since 1949. The same was true of the "homogeneous reactor experiment": an advanced prototype was to be constructed as an extension to a program underway at Oak Ridge which had demonstrated circulation of fuel, moderator, and coolant together through a heat exchanger.

The other two projects were demand-determined, in the sense that they were attempts to produce new reactor types even where there were no existing experiments. A small-sized reactor with appreciable power was required by the military for submarine development. Regardless of limitations in data then available in thermal core physics, a "submarine thermal reactor" was to be constructed because slow neutrons were easier to control in that environment. The last reactor — an operating core and heat transfer mechanism to test materials — was needed to overcome materials stress and corrosion in the other three projects. It was derivative from the other three, but only in specifics of the design since a general test facility would have been constructed for most other possible projects. The four projects comprised an *ad hoc* plan: choose reactor concepts that are ahead in parallel experiments in the national laboratories, fill-in between these projects in materials sciences, and accede to military requirements.[2]

Research under the *ad hoc* plan was not the only work done in the early 1950's, however. There was an extensive program not related to specific reactor concepts but consisting of a large number of smaller experiments with fuels or capital components in simulated reactor environments. Coolant experiments with boiling water at the Argonne

[2] For more description of these projects, cf. U. M. Staebler, "Objectives and Summary of USAEC Civilian Power Reactor Programs," *The Economics of Nuclear Power*, edited by J. Gueron *et al.* (New York: McGraw-Hill, 1957).

National Laboratory, and studies at other national laboratories of radiation damage to metals and to other, more erratic, coolants provided basic knowledge for test reactors but was not specific to any project begun at that time.[3]

In February, 1954 a formal plan was set up to make additions to and subtractions from these *ad hoc* Atomic Energy Commission projects. Research and development had to pass through three stages: (1) problem solving in basic physics or metallic alloys, (2) the construction and operation of "first-of-a-kind" or prototype reactors to show the technical feasibility of systems, (3) the construction of demonstration reactors to generate electric power at costs comparable to those from alternative fuel sources. The first two stages were to be in the province of the Atomic Energy Commission with Federal support of both national and private laboratories for the accomplishment of the basic and applied research, and with Federal expenditures for the development and construction of the first-of-a-kind reactors. The private companies that manufactured reactor components or that produced electricity participated at the second stage and provided most of the third stage. These corporations took part in first-of-a-kind reactors through "buying the steam" — either by constructing the steam and electricity generating facilities of the system or actually purchasing steam from an AEC-owned reactor. The corporations were to construct the third stage reactors themselves with Atomic Energy Commission assistance of unspecified nature and extent.[4]

The plan itself was to build five first-of-a-kind reactors for pre-designated levels of costs. The five were first steps on divergent paths to economic power capacity — some based on forecasts of low capital equipment expenditures but an inability to utilize fuel so as to compensate for high uranium prices, while others were based on high capital costs but low fuel prices to produce the cheapest power. Each faced formidable technical problems which had to be solved before steam could be generated for a reasonable length of time.

The *fast breeder reactor* program, which began with EBR-I in 1951, was to be expanded with the construction of EBR-II in 1956–1958. This second reactor was substantially a "first-of-a-kind" because the increase in its size over EBR-I raised formidable new research problems.

[3] Cf. Staebler, "USAEC Civilian Power Reactor Programs."

[4] Staebler states that industrial power reactors at the third stage were to be "fully justified on an economic basis" which would imply no need for Atomic Energy Commission assistance; but this stage was part of the Atomic Energy Commission program, so some assistance must have been intended. Since no reactors went this far, these financing questions did not arise. Cf. Staebler, "USAEC Civilian Power Reactor Programs."

Capacity was increased from 1,400 kW_t in EBR-I to 62,500 kW_t,[5] and the internal temperature of the primary heat transfer loops was raised from 600°F to 1,000°F. The first was not a simple task given that the larger and more numerous components had to be of perfect tolerance to prevent fouling and leak proof to prevent explosion of the liquid sodium coolant on contact with water. The higher temperature was necessary to attain the capacity/fuel ratios required for long run economic feasibility, but the resulting level of neutron acceleration was so high that the fission reaction sequence was difficult to control;[6] at the least, physics research predicting the response patterns to temperature changes remained to be done.

The *homogeneous reactor* program was also to be expanded, and in a manner quite similar to that for fast breeders. The Oak Ridge research group had constructed HRE-1 in 1952 under the first Atomic Energy Commission reactor program, and had operated it for almost two years before the 1954 five-part plan was under consideration. There was sufficient experience to show that fuel, moderator, and coolant could be combined and circulated through the coolant loop while still containing most of the fission in the core.[7] That is, a primary coolant loop and heat exchanger could be built which was tight enough to prevent leakage of radiation throughout the reactor. But the question whether the primary system pressure could be raised from 1,000 psig in HRE-1 to 2,000 psig, a level more in keeping with efficient fuel utilization, remained unanswered by tests in HRE-1. The development of coolant pumps and leak-proof components in a higher pressure system required experiments and tests of mock-ups similar to those used previously at Oak Ridge but of another order of magnitude in expenditure on materials and personnel. Corrosion and radiation damage still dis-

[5] The thermal capacity of the system is the total quantity of heat, in hundreds of calories per kilogram, in the steam per unit time. This is equivalent to the number of kilowatts of electrical capacity in the system if each calorie of heat can be completely utilized in the turbine generator without (a) loss before entering the generator, or (b) rejection of heat by the turbine to the condenser. Of course, this is not the case. If the efficiency of the system is, for example, 35 per cent, then the electrical kilowatts equal thermal kilowatts times 35 per cent. Cf. for example, G. Yekholodovskii, *The Principles of Power Generation* (New York: Pergamon Press, 1965), p. 91, *et seq.*

[6] The general complaint was that if the sequence from neutron to atom to neutron was disturbed by the temperature surge which increased neutron acceleration, then more fission would occur than previously. This would raise the temperature in the reactor core, which again would increase fission, so that any deviation from the original sequence, from random temperature variations, would not dampen down. This problem, while it did not arise in fact in the question of EBR-II, determined the course of the research in fast breeder reactors in the 1960's.

[7] The core consisted of little more than a "bulge" in the primary loop large enough to sustain fission.

rupted the operation of the first HRE; the circulation of uranium sulfate at 2,000 psig magnified the possibilities of extended disruption so that research had not only to continue in order to solve old problems but to begin anew with new materials in anticipation that the old problems would be much greater in the new environment.

The submarine reactor program had produced a working system for generating steam from fission. Distilled water under pressures of more than 2,000 psig, used as both moderator and coolant, proved to have good heat-transfer properties and to provide sufficient moderation for generating built-in limits on the rate of fission reactions. The system thus "worked" both in the sense of transferring heat from fission to a steam turbine according to design and in being able to return to design fission behavior if a random temperature or pressure change took place. Then the goal was to improve performance — to lengthen the lifetime of the fuel and reduce the amount of fuel required for a thermal kilowatt of capacity so as to make this system comparable in cost to fossil-fueled steam generators. Toward this result, a *pressurized water reactor* (PWR) program was established to build nonsubmarine prototypes. The new program was to rely on the results achieved so far with the submarine program, but to go further to "demonstrate that relatively large pressure vessels can be built according to specifications required for reactor operation . . . and (develop and) control very closely spaced fuel elements.[8]

The pressurized water reactor cycled H_2O under high pressure in liquid form; one alternative was to run at lower pressure and circulate steam through the reactor directly to the turbine generator. This was an attractive possibility because it eliminated much of the capital equipment for the pressure vessel (which might be 6 to 8 in. thick for the higher pressure PWR) and because it eliminated the equipment in the heat exchanger entirely. But development was no more complete than for the homogeneous reactor, nor as complete as for the pressurized water reactor in the submarine program. No working system had been constructed, even at the experimental stage, so that the first *boiling water reactor* had to be an experiment to "determine whether it can be operated without significant deposit of radioactivity in the turbine, the condenser, and the feedwater pumps (from the steam leaving the fuel core)."[9] There was no expectation that this project would lead directly

[8] *Abstract of Unclassified Material from Classified AEC Report to the Joint Committee on Atomic Energy*, "Program Proposed for Developing Nuclear Power Plant Technology" [received by the Joint Committee on Atomic Energy, March 12, 1954; cf. *Report of the Subcommittee on Research and Development on the Five-Year Power Reactor Development Program Proposed by the Atomic Energy Commission* (Washington, D.C.: U.S. Government Printing Office, 1954)].

[9] *Ibid.*

to a demonstration plant: "EBWR (Experimental Boiling Water Reactor) is not expected to provide nuclear data on the critical mass of a large reactor or on proper spacing of its fuel."[10] Another stage of development beyond that planned in this five-year program was required before blueprints for economic boiling water reactors could be given out.

The fifth project in the plan was to build a reactor experiment with liquid metal (sodium) as the coolant and graphite as the moderating material. The *sodium graphite reactor* was not a direct descendant of the early postwar experiments in the national laboratories but rather was cross-bred from a number of such experiments. The breeder and other sodium research had demonstrated the heat-transfer capability of sodium (as well as the problems to be encountered in heat exchanger leaks from sodium burning in air or water), and small-scale experiments with graphite in the national laboratories had shown high quality performance as moderating material. Independent design studies of sodium graphite systems (by Atomics International Company) were added to the experimental work; the sum total of promising results and designs led to the decision to take on this reactor concept in 1952. There was much to be accomplished by building an experimental plant: "Many features of the plant, and its operating procedures, have not been tested in reactor practice . . . the upper limits for fuel and coolant temperatures, burn-up and other operating variables . . . moderate changes in these variables such as increasing maximum uranium metal temperature from 1,200°F to 1,400°F and maximum coolant temperature from 1,000°F to 1,250°F could have an appreciable [reducing] effect on the cost of power."[11]

The final goal of all five reactor programs was to demonstrate the technology necessary to build steam generating capacity at $50 to $70 per thermal kilowatt (or $200 per electrical kilowatt). There was widespread confidence that the relative abundance of uranium over fossil fuels such as petroleum and natural gas would make nuclear fuel less expensive; but there was less confidence that nuclear capital costs would be comparable to those for the much simpler fossil-fuel-burning boilers: "The problem of developing nuclear reactors for the economic generation of electrical power is largely one of reducing costs for capital investment . . . for practical nuclear power plants of the future, construction costs of $50 to $70 per kilowatt of heat are sought."[12]

The goal was not attainable within this program by each of the five research groups building reactors. The pressurized water program called

[10] *Ibid.*
[11] *Ibid.*
[12] *Ibid.*

for a large reactor (264 MW_t)[13] at Shippingport, Pennsylvania, which would be the first "full-scale" plant and would be expected to demonstrate that one additional round of component improvements would result in economic nuclear power. The four other programs were frankly experimental; they had to show that reactor concepts could be built and could be operated at least 80 per cent of the time after construction was complete.

The first project was expected to cost $800 per electrical kilowatt,[14] while the others were to cost from $600 to $1,000 per kilowatt. They were constructed according to the experience shown in Table A.1.

Table A.1 The Reactors in the 1954 Five-Year Program

Reactor Type	Plant	First Predicted Costs	Actual Costs
Fast Breeder	EBR-II	$740/kW (1957)	$2,320/kW (1961)
Homogeneous	HRE-2	$1,050/kW (1956)	$1,650/kW (1958)
Pressurized Water	Shippingport	$800/kW (1957)	$1,250/kW (1957)
Boiling Water	EBWR	$900/kW (1957)	$1,180/kW (1957)
Sodium Graphite	SRE	$630/kW (1956)	$960/kW (1957)

SOURCE: Atomic Energy Commission reports to Congress on the Civilian Nuclear Power Program (most frequently before the Joint Committee on Atomic Energy, Hearings pursuant to Section 202 of the Atomic Energy Act of 1954, from 1960 to 1967).

Shippingport proved to be operational at 50 per cent more capital expenditure than called for in the original design, but at about the time originally planned. The four experimental reactors had technical difficulties of different degrees of seriousness. No single experiment was constructed with capital costs equal to the original design expenditure; and the total expenditure averaged almost twice that for the original design. All took somewhat longer than expected to put into operation: one was completed late in the year originally forecast; two were one or more years late, and the last was completed four years behind schedule.

[13] A thermal megawatt is 1,000 kW_t or 1,000,000 W_t.

[14] The capital costs on this page and those following in this appendix are quoted in terms of dollars per electrical kilowatt — equal to total expenditure on all items of capital, on interest charged during construction, and on research and development expenditures specific to this plant, all divided by the total net electrical capacity of the plant. This departs from the emphasis in the preceding pages on production of heat energy or thermal kilowatts. The capital costs per unit of heat energy can be estimated by multiplying these statistics by plant efficiency, usually of the order of 30 to 40 per cent.

The five-year plan was put into effect and its goals were realized to a limited extent. Shippingport proved the point it was supposed to prove, in part: although it did not demonstrate power capacity at $800 per kilowatt (including expenditures on original research), it showed well enough that the next round could take place at a lower level of costs. Two private PWR power reactor projects were announced subsequent to the first few months' operation, and both had forecast capital costs in the $200 to $300 per kilowatt range. There was no assurance from Shippingport's experience that these costs would be attained, or that fuel costs were "negligible"; but the broad outlines of cause and effect from prototype to demonstration were being followed. The experimental projects were successful to a more limited degree. Three of them resulted in working reactors in the five years of development, albeit at higher cost levels than expected; but one had not yet begun operation by 1958, so that only three out of four had "proven out" within the standards set by this program.

Planning and strategy are difficult to tell apart at this very preliminary stage in reactor development. The five-year plan set out the goals for three stages of development, and also set out the five projects to attain either the first or second stages, without hinting at changes which would take place if the first-stage goals were not reached. The expenditure goals were not achieved nor were scheduled dates of completion generally met, but projects were not curtailed. Then the plan must have been the strategy: the five projects were to be continued, even if there were significant departures from expected results.

Strategy and the 1958 Ten-Year Program

A draft statement of a long-range plan for the development of civilian nuclear power was published by the Joint Committee on Atomic Energy in August, 1958.[15] The goals of the program, to be reached before 1969, were centered on producing "competitive power in high cost energy areas." Both goals and the means for achieving them were specified in some detail; they showed a tendency for Atomic Energy Commission program planners to move further in the directions already laid down in the previous five-year program.

Competitive power from a reactor system was said to be available when "utility executives [could] make a decision to build nuclear stations on the basis of lower costs over the life of the reactor."[16] Not

[15] Cf. *Proposed Expanded Civilian Nuclear Power Program* (Washington, D.C.: U.S. Government Printing Office, August, 1958).

[16] John A. McCone, testimony before the Joint Committee on Atomic Energy, *Hearings Pursuant to Section 202* (February 16, 1960; pp. 122, *et seq.*).

all utility companies were expected to be able to make such a comparison, but rather only those in "high cost energy areas" where electricity from fossil fuel systems was "estimated to be [available at] 7 mills/kWh."[17] These costs were to be met by combinations of nuclear fuel expenditures of 2–4 mills/kWh and capital expenditures in the range of $100 to $300 per kilowatt of capacity. The exact target depended on assumptions on the percentage of time that any one reactor system would operate, on prices per unit of fuel and of capital, and on interest charges; the approximate target was to undercut fossil fuel costs by more than half, so as to make it possible for nuclear capital costs of $200 per kilowatt to be competitive. This target reaffirmed that of the 1954–1958 program, as far as capital costs were concerned; but it extended beyond that in the earlier plan by making explicit the necessity to reach fuel costs as low as 2 mills/kWh.

The research projects were also extensions of those in the first program. A number of routes were to be taken to demonstration reactors — all beginning with experiments on reactor components, continuing with construction of "first-of-a-kind" small-scale reactors, ending with construction and operation of full-scale plants showing the attainment of target costs. But there were to be more such steps in this sequence and more reactor types at each step.

First-of-a-kind reactors were to be built in two sizes, one at a small scale for experiments with original components and newly developed materials in simple systems (termed "experimental power plants"), the second at full scale for collection of information on the behavior of these components at useful size (termed "prototype power plants"). The second-sized experiment was an additional step in development, given that "advanced prototypes" and "demonstration reactors" completed a project.

More reactor types were to be made the subjects of research than ever before. The original five reactor programs were to be continued, with the four preliminary programs going to larger prototypes, and the advanced program to both an additional, very advanced prototype and to demonstration reactors. Large-scale prototypes were to be added to the *fast breeder*, the *homogeneous*, the *boiling water* (where three new prototypes of various sizes were scheduled), and the *sodium graphite* reactor projects. The prototype plant at Shippingport was not to be terminal development in the *pressurized water* reactor project, but was to be supplemented with three newer reactors which would attempt to improve performance, mostly by adding to steam temperatures and pressures. Three new projects for new reactor types were added to the

[17] *Ibid.*

Table A.2 Forecast and Realized Costs for the 1958 Ten-Year Program

Reactor Type	Ratio of Realized to Forecast Expenditures on Construction, for all Reactors of that Type Completed Before 1967	Number of Reactors
Pressurized Water Reactor	1.2	3
Boiling Water Reactor	1.4	5
Heavy Water Reactor	1.7	1
Gas-Cooled Reactor	1.9	1
Sodium Graphite Reactor	2.4	1
	>2.0	Canceled
Homogeneous Reactor	1.6	1
	Unknown	Canceled
Organic-Cooled Reactor	2.1	1

SOURCE: Various fiscal reviews of the civilian power program, from annual publications of the Atomic Energy Commission and the Congressional Joint Committee on Atomic Energy.

NOTE: The estimates for the Fast Breeder Reactor are not shown, for reasons given in footnote 20 of this appendix.

original five at a level of research activity quite comparable to them: one project using an *organic coolant* (which had been operated at a lower level for some time, outside the five-year program); a second project for a reactor using *heavy water* as both coolant and moderator (based mostly on promising experiments in Canada); and the *gas-cooled reactor* project, expanded from what had been a number of private ventures with helium cooling by one of the private reactor manufacturers. The entirely new projects were to start up with small experiments, and to continue through to the final demonstration reactor. By Congressional count, "present plans call for design studies during the next five to seven years of 21 reactors, including nine of large size, four of intermediate size, three of small size, and five small experiments."[18] Not all of these designs were expected to become working plans for plants, but 14 were to be built for expenditures of $0.4 billion for plant and equipment and of $1.3 billion for general research.

The strategy for effecting the plan consisted of some working rules as to choice of actual projects when the time had arrived to begin construction. There were more designs than construction projects: "about half of the most promising designs would be carried through the construction stage . . . (with) total capacity of plants to be built in this

[18] Cf. "Comments of Reactor Designers on the Proposed Expanded Civilian Nuclear Power Program," Joint Committee on Atomic Energy, 85th Congress, December, 1958.

country . . . expected to amount to about one million electrical kilowatts."[19] The choice of few plants from many designs was to be made mostly within reactor types — for example, four designs of a reactor type might precede the construction of one small-scale experiment of that type — but implicit in the program was some comparison across types, and a mandate to eliminate any of the parallel projects that fell considerably behind the others.

The strategy actually put into effect over the 10-year period was to cancel the advanced steps in any project that appeared to be "far behind." All projects failed to stay within predesignated levels of capital costs; as shown in Table A.2, the average ratio of actual to design costs was greater than 1.5 for all projects. But those coming out of the previous five-year program at a fairly advanced stage of development — beyond the small-scale experiment stage into that of construction of large-scale prototypes[20] — can be divided into three classes. The first class projects showed costs 20 to 40 per cent greater on average than expected, but a tendency to approximate design levels of costs more closely in later reactors. These projects, the two *light-water* projects above the first line in Table A.2, were advanced through each of the designated stages until they achieved the program target of demonstration reactors costing $200 per kilowatt of capacity. The second class projects showed costs 70 to 90 per cent greater than prescribed at the stage of construction of fairly advanced experiments; the absolute levels of these costs were very high as well, since they exceeded $1,350 per kilowatt. The two reactor types in this class, the *heavy-water* and *gas-cooled* projects below the first line in Table A.2, were put on a "contingency" basis by making later stages of development depend on and wait on the performance of the two experimental reactors which were under construction in the early 1960's. The third class projects were those that were canceled. Two of them, the *sodium graphite* and *organic-cooled* projects, had shown costs twice as much as expected; while the *homogeneous* reactor experiment had costs only 1.6 as much as originally forecast, it was able to provide capacity only at $1,650 per kilowatt. None of these in the last group had provoked private developers into

[19] *Ibid.*

[20] The fast breeder program was still at a much earlier stage of development. Although small experimental reactors were in use or being constructed (such as EBR-I and EBR-II) and a full-sized prototype was under construction, standards for reactor operation were different from those applied to other projects. The time-length of development was to be much longer than for other reactor types, because the optimal time for installing breeders appeared to be in the late 1970's or early 1980's. The expected costs of the program were much more extended, as well, and those incurred at this stage were only preliminary to a much larger program. As a result, immediate availability of a working economic reactor was not required for this program.

plans for cooperative prototypes, because other projects already had lower forecast energy costs at the same or a more advanced stage of development. These were reasons for cancellation given by Mr. Milton Shaw, Director of the Atomic Energy Commission's Division of Reactor Development, in his discussion of the *sodium graphite* project: "The utility interest is not there; the economics are not there in terms of positive factual evidence to date . . . the resources put into this plant thus far in terms of sodium technology could better be applied to the liquid metal fast breeder program."[21]

The strategy was in fact to choose from the 10-year program those particular types that had lower forecast energy costs and the better record of cost performance on construction of research reactors. This was a matter of fact, since from among eight projects — all going through two of four steps to demonstration reactors — two thermal projects were financed to the final stages because they had managed at the early stages to pull ahead of the others. These two had quite similar construction costs, so that the choice between them was of second-order importance; they had costs per kilowatt of capacity which came closest to design costs, and they had lower construction costs at each particular step in development.

The Results of the 10-Year Program

By the standards of measurement set up in 1958, the 10-year program was a success by 1967. Two of the eight reactor projects had achieved the goals set for capacity costs — more than achieved these goals, since capital expenditures in new designs for *light-water* reactors had fallen considerably below the $200 per kilowatt that was sought. The development pattern was most striking and quite similar in the two reactor projects; both showed significant progress quite early in their respective programs.

The *pressurized water* reactor, which had been shown to be operable in the Shippingport case, displayed rapid and substantial improvement in cost and operating performance in the experience with the Yankee Atomic Electric Plant. This plant, put into operation at Rowe, Massachusetts, in the middle of 1961, showed completely reliable production of electricity from that point on; except for refueling, the plant remained in full operation and after the first refueling its rated capacity was raised 10 MW beyond the 155 MW expected. The Yankee design called for more capital expenditures than were actually made, and actual capital

[21] Testimony of M. Shaw, *Hearings Pursuant to the AEC Authorizing Legislation Before the Joint Committee on Atomic Energy*, February–March, 1966, Part II, p. 819.

costs were close to $300 per kilowatt. This was a decrease from Shippingport of more than $900 per kilowatt, so that the Shippingport experience along with further research expenditures of $5 million in Yankee had brought the project more than 80 per cent of the distance to the $200 per kilowatt goal. Three additional reactors of the *pressurized water* type were undertaken by private developers with Atomic Energy Commission research grants[22] around the time that Yankee began operation; even though not all of them have yet been finished, they proved to be the last round of research. More than 15 private projects were initiated in the 1964–1967 period with scheduled construction expenditures considerably less than $200 per kilowatt of capacity.

The outlay on this project, and the results therefrom, can be characterized by taking a macro view of the construction of experimental, prototype, demonstration, and operating electricity generating reactors. This view of research is that, for a year's research outlay, a year's improvement in performance is obtained: $\pi_t = f(E_t)$, where π_t is a performance index on which to measure research progress, and E_t is an expenditure index, both occurring in year t. For the *pressurized water* reactor project, π is design capacity costs of $200 divided by the lowest costs actually achieved on a working PWR in each of the years 1957–1965, and E is the per cent the research expenditure in each year is of total 1957–1965 PWR expenditure. The function $\pi = f(E)$ that fits the nine years of experience best is the least-squares equation $\log \pi = -0.249 + 1.28 \log E$, with the coefficient of correlation $R^2 = 0.876$; it indicates that results increase at a faster pace than expenditures (since the coefficient showing $d \log \pi / d \log E$, or percentage change in π/percentage change in E, is $+1.28$). This may be somewhat misleading, since a great part of the early experimental PWR work was done under the aegis of the submarine thermal reactor program and is not included in E; but as a first indication it would appear that the appreciable progress already made at the beginning of the project increased at a substantial rate with only small additional expenditures.

The successful completion of the project to prove out a *boiling water* reactor was not achieved with the assistance of basic research from a military program. But the similarities of results in this case and the accomplishments of the PWR program are still quite strong. The large step for the PWR project in the installation of Yankee Atomic was paralleled for the PWR project by private construction of Dresden 1

[22] The terms of modified "third round" assistance were that, if the companies built and operated the reactor systems, the Atomic Energy Commission would contribute an amount to be negotiated for research and development expenses, and would provide the fuel core without charge. This must have placed the impetus on reducing capital charges rather than fuel costs since — in effect — other inputs besides capital were available without cost.

at almost the same time. Dresden showed that the *boiling water* concept could be put into effect for reliable power production — in 1964, for example, the plant had a forced outage rate of only 2.2 per cent of total available operating time — but the costs of construction were planned at $200 per kilowatt and realized at $340 per kilowatt. A major round of advanced development was still required, if the target of $200 per kilowatt was to be reached.

The approach to advanced development differed somewhat from that in other projects. Three prototypes were constructed as joint AEC–private utility research reactors, all in the range of 20 to 70 MW of capacity, each to center attention on solving a different problem. The Elk River reactor, with a closed cycle of water coolant from reactor core to turbogenerator and back, was completed in 1962; the Big Rock Point reactor, to demonstrate operation of new types of fuel with extremely high thermal energy per unit volume, was completed the same year. They provided new and important leads to reducing the costs of the heat transfer system and fuel burn-up, respectively. But these two plants did not themselves show lower capacity costs than in Dresden 1; their contributions lay in providing better components or fuel for later reactors, so that little was achieved immediately from large additional expenditures for research.[23] The next round of construction after this achieved the target, without any further research expenditures: plants at Nine Mile Point and Oyster Creek were ordered for 1967 with costs of construction of less than $200 per kilowatt, and were accompanied by announcements of more than a dozen orders for BWR plans of 500 to 1,065 MW capacity by private utilities in the 1965–1967 period.

This experience with research prototypes strongly affected the pattern of results per dollar of expenditure. For the least-squares log regression, $\log \pi = -0.250 + 0.531 \log E$ and $R^2 = 0.622$; the curve does not "fit" the nine years of observations of percentage results π and percentage expenditures E very well, since the coefficient of correlation is only 0.622 for a very small number of observations. But it indicates as a first approximation that diminishing returns to research expenditures were predominant — on average, an additional percentage point of expenditure brought the project 0.531 of a percentage point closer to target that year — and this can only be attributed to the large outlays with no reductions in capacity costs during the middle years of the program.

[23] The third prototype, the LaCrosse BWR, was constructed to carry on experiments in the use of mild alloy steels and to develop advanced steam separators. It was not completed until 1967, so that it contributed little to the ten-year program discussed here.

The two reactor projects "on target" should not be greatly differentiated because of special circumstances in each. The general description of a successful project might well center on $\pi = f(E)$ for 18 observations, nine for the PWR and nine for the BWR in the years 1957–1965, so that the least-squares equation that fits the combined experience is $\log \pi = -0.327 + 0.731 \log E$ with $R^2 = 0.557$. This derived relation fits better than that for the nine observations of the BWR, while it is no worse a fit than that for the PWR, so that there is some net gain from putting the two sets of observations together.[24] This summary view shows diminishing returns to research expenditure, since the average movement toward the target was only 73 per cent of the annual additions to expenditures (that is, $d \log \pi / d \log E = 0.731$). Even the successful projects experienced declining relative results soon after the experimental mock-ups were completed.

Lessons for Strategy

The success achieved in the two water reactor projects, and the contrasting lapses experienced in four other projects, would seem to be the source of two important pieces of practical wisdom. The first has to do with deciding on the numbers of research projects, when the important consideration is the total amount of research expenditure, and it prejudices the case toward more projects rather than fewer. The second is pertinent to the same decision, but in another aspect: attaining the greatest possible spread of benefits from nuclear research to the consumers of electricity. This finding also favors the larger number rather than the smaller number of projects.

There is first some indication that more research projects can be cheaper than fewer research projects, "per unit" of research results. This follows from two aspects of the experience in the 10-year program: all prototype construction projects exceeded cost forecasts, and not all projects achieved the program target after reasonable additional expenditures beyond those originally expected. The average expenditure on those reactors actually constructed in all of the projects was 180 per cent of the original amount specified. The additions made to the two light-water projects were enough to reach the $200 per kilowatt goal set for private construction costs, and further "unexpected" expenditure might have brought the other reactor types to this goal as well.

[24] The test here for "goodness" of fit is simple and direct: assume that the correlation coefficient is different from zero as a result of chance (so that there is no fit); if the chance of this happening is less than 2/100 abandon this assumption for the contrary that there is a significant fit. The first assumption cannot be abandoned, on these standards, for the BWR; it can for the PWR and the combined sample of BWR and PWR observations. Cf. Table V-A in R. A. Fisher, *Statistical Methods for Research Workers* (New York: Hafner, 1958).

The question is "how much" further expenditure. One conceivable answer follows from assuming that these other reactor types could have achieved the target by following the research pattern $\pi = f(E)$ set by the two successful projects, and that they were at E_0 implied by realized π_0 at the time they were downgraded. This is to say that all projects were on the path $\{\log \pi = -0.321 + 0.731 \log E\}$, but that the five lapsed projects were at the wrong points on the path; for one thing, the sodium graphite project was observed to be at $\pi_0 = 0.20$ and $E = 0.82$ in 1963, but — for that level of achievement — was actually at $E_0 = 0.41$ of total expenditure for a *successful* project.[25] Then the downgraded projects would have required expenditures in the mid-1960's roughly E/E_0 times those originally expected, in order to arrive at a point on the path $\pi = f(E)$ to successful research results. These expenditures would have been necessary to build successful prototypes; since each prototype probably would have cost more than expected — as had been the case on all projects — then the total cost of taking any one project to completion, as a ratio of original expected outlays, would have been the product of E/E_0 and the realized construction costs on each reactor shown in Table A.2.

The relevant choice in retrospect was between a strategy of selection of promising projects from a number of parallel projects, and a strategy of "sticking to one's last" with a single reactor type.

The results from the first strategy are indicated by actual events: two successful design types for the costs of close to six projects (the expenditures on the five downgraded projects came to 4.4 times the average of the original forecast expenditures per project; with the light-water reactor costs, an outlay equivalent to that for six projects was incurred).[26]

The results from the second strategy would have varied from case to case, depending on which single reactor system was chosen before the 10-year program began. If the PWR had been chosen, then one successful design would have been attained for 1.2 times the original forecast costs; on the other hand, if the homogeneous reactor concept had been chosen, it seems most probable that the costs of success would have been more than seven times those originally forecast for that single project alone [4.5 times the original costs, so as to move onto the successful growth path $\pi = f(E)$, multiplied by 1.6 times the forecast costs for each reactor of this type constructed]. It might be assumed that the promise of success was about the same for each reactor type before 1960; although there was a military PWR reactor working, the chances of *economic* success with that concept might have

[25] That is, for $\log \pi_0 = \log(0.20) = -0.321 + 0.731 \log E_0$, then $E_0 = 0.41$.

[26] This once again ignores expenditures on fast breeder reactors since it is assumed that the results of the fast breeder project are not yet known.

seemed no greater at the time than with the six other types not yet to the prototype stage. Then the results from the single project strategy would have been the average of results from pushing each of the separate types to a final conclusion (where each of the separate results was E/E_0 times the ratio of realized to forecast construction costs). This average after the fact is close to 3.5 times original forecast costs, so that it would have taken the cost equivalent of 3.5 original projects to prove out one concept by centering all attention on that one concept.[27] Then in order to develop two reactor concepts, (1) start out with six concepts and discontinue all except the two best after some experimental results with prototypes for each; (2) start out with only two concepts, and spend an amount *on average* equivalent to forecast costs for seven projects before these two reach the target.

The second lesson is that more projects mean more companies producing reactor systems, and this might well result in faster diffusion of the fruits of innovation. The first part of this lesson is the easier to believe, because the results of the 10-year program showed that the manufacturer constructing the experimental and advanced prototypes for a particular reactor type ended up with almost all of the private industry orders for that type. Westinghouse constructed Shippingport, then Yankee, along with all of the advanced reactors that followed this important second plant; all except 5 of the 26 private *pressurized water* reactors subsequently on order through March, 1967 were on contract to this company.[28] The General Electric Company was responsible for the construction of the Dresden and Big Rock Point prototypes that made up the first and second steps in the *boiling water* reactor project, and this company completed most of the remaining prototypes as well.[29] General Electric contracted for 20 of the 23 *boiling water* reactors on order through the first three months of 1967, a substantial lead in sales of this concept (marginally contested only by Combustion Engineering).

Such results are not surprising. The achievement of a reactor design goal results from construction of reactor systems — from "learning by doing" often enough to gain control of the interaction of new com-

[27] An alternative assumption is that the subjective probability of final success was twice as great for the PWR than for any other design type before the fact. This does not change the estimates: the costs (weighted by probabilities) averaged 3.2 times those forecast, rather than 3.5, so that it might have been possible to do two projects successfully for the forecast costs of 6.4 rather than 7.0 projects.

[28] The others were contracted in 1966 and 1967 by Babcock and Wilcox, the producers of the one other PWR prototype constructed in the early 1960's; the exception might well demonstrate the rule.

[29] Allis Chalmers constructed the important Elk River Reactor, but left the industry in the late 1960's; Combustion Engineering entered only at the last stages of prototype construction with BONUS which was scheduled for completion in 1968.

ponents and to improve on these interactions. Success was measured in terms of the operating performance of a reactor costing less than $200 per kilowatt, and if there were no prototypes, then no measure on this standard could be given. Grades on the first prototypes built by a company were invariably lower than on second or third prototypes of other companies built at the same time.[30] Then the observed relation between π and E for the two light water reactors was not only a particular path to economic capital costs, but also the "learning curves" for two separate corporations. At the end of the two projects, each had a decided advantage for sales of one or the other reactor type because they could and had built enough reactors to control costs and system performance.

The last part of the lesson is that the existence of more corporations in the market implies more competition and thus a faster spread of the benefits of research to the consumer. This is not always the case; a larger number of competitors is not synonymous with market behavior characterized as "competitive." Three companies, all producing turbine generators, operated an organization for price setting in the 1950's that had all of the effects of monopoly; three was not a large enough number for competitive behavior to result from the price-output policies of firms.[31] On the other hand, two independent firms in other industries can and have produced price-cost differences, price flexibility, and lack of discrimination between buyers that can only be characterized as "competitive."[32] In the case of nuclear reactors, the pattern of output and price formation does seem to be such that the greater the number of research projects, and thus the greater the number of firms, the more extensive the competition.

The sales of reactors to private and government utilities began in earnest in 1965, but for delivery in 1970 or thereafter.[33] General Electric agreed to construct 5 of the 9 for delivery to domestic firms in 1970, 6 of the 12 for 1971, but only 1 of the 12 for 1972; Westinghouse was responsible for 3 of the 9 for 1970, 3 of the 12 for 1971, but 7 of the 12

[30] For example, the first Allis Chalmers BWR had capital costs $60 per kilowatt higher than the second General Electric BWR finished in the same year. At a later point, the lead had changed so that the second Allis Chalmers experiment cost $800 per kilowatt less than the first Combustion Engineering BWR experiment. (These indicate no more than that the number of previous experiments, as well as the size of the reactor and the difficulty of the experiment, might be a factor in determining the costs of construction for the prototype.)

[31] Cf. Testimony of M. A. Adelman in *OVEC* vs *General Electric*, Docket 62 Civil 695, p. 4900 *et seq*.

[32] Cf. James McKie, *Tin Cans and Tin Plate* (Cambridge, Mass.: Harvard University Press, 1959), Ch. IX.

[33] Cf. Joint Committee on Atomic Energy, *AEC Authorizing Legislation — 1968* (Washington, D.C.: U.S. Government Printing Office, 1967), pp. 707–709.

for 1972 (with the residual supplied by Babcock and Wilcox, or by Combustion Engineering). The first company altogether accounted for 12.5 thousand megawatts of capacity on order, the second for 11.7 thousand megawatts, of a total of 31.3 thousand; each had exactly 20 plants under contract as of the third quarter of 1967.

Two aspects of this order pattern stand out rather clearly. The market for new reactors has been shared in a remarkably even fashion so far, 39 per cent for General Electric and 38 per cent for Westinghouse. Also, the sequence of orders is quite separated in time, with General Electric "filling up" first while Westinghouse has taken almost all of the orders for later delivery. Both might be considered the results of a market-sharing agreement between these firms; but there is no evidence of such an agreement, nor is it likely to have been in effect given the surveillance of these companies by the Justice Department and the buyers after the "great electrical conspiracy" was revealed in the early 1960's. Both aspects of behavior are consistent, as well, with a contrary hypothesis — that each company ignores the output behavior of the other in the "Cournot" manner.[34] Each producer regarding himself as the single seller of a *differentiated* reactor concept — one unique in enough aspects of performance to make demand separate from that of other reactor types — would treat the demands of utilities as dependent only on his quoted charges for construction of the system. This would be a mistake, since the other companies' charges would in fact be almost as important, but the result of the mistake would be to set demand to be equal to total reactor demand minus the average sales of the other firm. The two companies following this output strategy would each make about the same number of sales as the other; most likely, when one company started selling slightly before the other, each would make his sales "in turn."[35]

The implications of Cournot output behavior extend beyond equality of shares to a particular relation between the level of prices and the number of firms. When there is one firm, Cournot pricing is monopoly pricing; this price might be termed P and output termed q. When there are two manufacturers, as a result of the entry of a second firm, the new firm considers q as fixed and chooses an output q^* for himself; this additional output causes the first firm to cut back on his original production to q^{**}, because $q + q^*$ reduces price under that for q alone. The roles of each are taken by the other over again, until $q^{**} = q^*$ and $q^{**} + q^* = \frac{3}{2}q$. The result of the entry of a second firm is an

[34] After A. Cournot, *Recherches sur les Principes Mathématiques de la Théorie des Richesses* (1838).

[35] Where the limit on sales by the first firm is set by short-term capacity.

increase in market output and consequently a reduction in profit margins to one half the previous, so as to "pass through" the fruits of invention.[36]

Does the Cournot model describe the price behavior of the two light-water reactor producers? Price reduction took place after it was known that there would be two producers on equal terms, so that prices for two were less than for one. Certainly the General Electric price list was reduced by more than 20 per cent after the Westinghouse contracts in 1963 had shown that this other company was going to reach the $200 per kilowatt target, and the reduction took capacity prices well below $200. But there were quite obvious cost reductions for the two producers from learning to build reactors that explain these price reductions just as well as does the Cournot argument. The only indication that prices are more similar to the two-producer Cournot case than to monopoly is in the broad estimates of price-cost differences for these companies.

Prices had settled, on the basis of a dollar of constant purchasing power, "to around a little over a hundred,"[37] with a low of $95 per kilowatt and a high of $120 per kilowatt for capacity in the 600 to 900

[36] The general argument is that the producer "i" maximizes net receipts NR with

$$\mathrm{NR}_i = q_i P(Q) - C(q_i)$$

where q_i is that firm's output, P that firm's price (and all firms' uniform price, a function of total output Q) and $C(q_i)$ that firm's total costs. All q_j, $j \neq i$ are assumed to be fixed by firm i, after Cournot. Then for maximum NR_i, $\partial \mathrm{NR}_i/\partial q_i = 0$ which requires

$$P(Q) + q_i \frac{\partial P}{\partial Q}\frac{\partial Q}{\partial q_i} - \frac{\partial C}{\partial q_i} = 0$$

for all firms i. If there are n such firms, so that there are n equations, then

$$nP + \sum q_i \frac{\partial P}{\partial Q} - n\frac{\partial C}{\partial q_i} = 0$$

(for $\partial Q/\partial q_i = 1$ and $\partial C/\partial q_i$ the same for all firms). Then the simple decision rule for the firm's profit margin is

$$\left(P - \frac{\partial c}{\partial q}\right) = -\frac{1}{n}\cdot\frac{P}{\epsilon},$$

where ϵ is the elasticity of market demand (equal to $[(\sum q_i/P)(\partial P/\partial Q)]^{-1}$ in the previous expression). When there is one firm,

$$P - \frac{\partial c}{\partial q} = \frac{-P}{\epsilon};$$

the monopoly profit margin equals $-P/\epsilon$. When the number of firms is two, then the profit margin is half that of the monopolist. When the number is so large as to be beyond count, the profit margin — the price-marginal cost difference — is zero. This is shown quite simply and clearly in W. S. Vickrey, *Microstatistics* (New York: Harcourt, Brace & World, 1964), pp. 337–338.

[37] Cf. Remarks of Chairman Holifield, *AEC Authorizing Legislation, Fiscal Year 1967* (Washington, D.C.: U.S. Government Printing Office, 1966), p. 659.

MW range. All indications point to demand elasticity in the range of -1.03.[38] Marginal costs — the change in a reactor producers' costs from adding an additional order to his production list — are, of course, not publicly known outside of the plants of the two producers; but an "order-of-magnitude" calculation can be made of these costs by adding up engineers' estimates of components and materials costs for a 600 MW reactor.[39] This assumes that there are lower costs for the last contract than for all of the others, given that engineering design expenses and assembly costs are not included at all (implicitly, these activities were "free" to the last contractee because resources to provide them had already been invested). The calculation shows that marginal costs could not have been less than \$60 per kilowatt, and were most probably in the \$70 to \$80 range. Cournot pricing for two firms has price minus marginal costs $\{P - \text{MC}\}$ equal to price divided by twice the elasticity of demand $-P/2\epsilon$ (for number of firms $n = 2$) while the absence of any effect from the second firm results in $(P - \text{MC}) = -P/\epsilon$. The estimate $(120 - 70) \sim -120/2(-1.03)$ is in accordance with the two-firm Cournot description; price is \$50 per kilowatt too low, or marginal costs more than \$50 too high, for the second firm not to have affected the level of output and prices.

Summary

There is little more evidence to firmly establish either lesson. But, rather than maxims for all future decision-making, there are indications that money was probably spared by starting more parallel projects and that a second successful project reduced profit margins. The question is whether future large-scale research projects will not repeat this history only because it was not taken into account.

[38] The econometric estimate of this value is discussed in detail in Appendix C.

[39] The engineers' estimates are from the monographs in *Guide to Nuclear Power Cost Evaluation*, Volume 3, *Equipment Costs*, Kaiser Engineers Rept. TID-7025, Vol. 3 (March, 1963). They hold for the early 1960's; but it is assumed that later reductions on materials costs are canceled by general price inflation.

Appendix B

Notes on the Costs of Research

1. The focus of the light water reactor program on thermal fission left important gaps in the knowledge as to the effects of fast fission on the core environment, and as to the rate of fast fission itself. Studies are underway to provide more information in both areas. Capture cross sections — the frequency distribution of absorbed fission particles — of structural materials are being documented by observations from experiments in the energy range 1 keV to 1.5 MeV; and the resonance and scattering of particles in both fertile and fissile materials in the energy range from 1 keV to 3 MeV are being observed in the same experiments. The sources of research and the progress are shown in Table B.1.

Table B.1 Studies of Fission Processes

Neutron Energy Range	Major Contributing Facility	Present Rate of Progress
3–14 MeV	ORNL — 5 MeV van de Graaff	
	Aldermaston 6 MeV van de Graaff	
	ANL — 3–10 MeV neutron generator	Poor
0.3–3 MeV	ANL, ORNL, Harwell — 3 MeV van de Graaff	
	ORNL — 5 MeV van de Graaff	Very good
1–300 keV	RPI + GA linacs	
	ORNL linacs	Very poor
100 eV–1 keV	RPI, GA, Harwell, Socley linacs	
	Columbia cyclotron	Very good

2. The facilities to test core configurations are ZPR-III, VI, IX, and ZPPR. ZPR-III is a fast critical facility with a 1,500 liter capacity. It is the only facility currently approved to investigate cores with large amounts of plutonium. It has been used to perform critical experiments for EBR-I, EBR-II, Fermi, Rapsodie, and Sefor.

ZPR-VI and IX are critical facilities designed to investigate the physics of large (3,000 liter) dilute fast reactors. They will be used to verify computational techniques and cross sections for large fast reactors.

ZPPR, now under construction and expected to be in operation by late 1968 or early 1969, is designed to handle plutonium critical assemblies as large as 3,000 kg; it will be sufficiently large to reproduce to scale the core for a nuclear plant with 1,000 MW_e capacity. Hence it will be extremely useful throughout the LMFBR program.

3. The Atomic Energy Commission Authorization Hearings for Fiscal Year 1968 show the following forecasts for expenditures on reactor safety programs (in millions of dollars):

	Fiscal Year 1968	Fiscal Years 1972–1980
Reactor Safety Analysis	1.5*	1.5
Engineering Safeguards System:		
safeguard technology	4.0	—
plant application tests	2.3	—
standards, codes, specifications	2.3	—
Nuclear Safety:		
SPERT	2.1	2.0
power burst facility	2.8	1.2
chemical reactions	2.5*	2.5
other reactor kinetics (two-phase flow in Na)	4.0*	4.0
Effluent Control R & D	4.7	
Engineering Field Tests	8.8	—
	35.0	11.2

* Indicates funds related more or less totally to fast breeder programs; the total is $8,000,000.

The statistics for 1972–1980 are estimates consistent with carrying on the 1968 program and with making extensions necessary for the expanded fast breeder program outlined here.

Appendix C

Documentation of Capacity Demand Equations and Total Capacity Forecasts

W. A. Stull and P. W. MacAvoy

This paper is concerned with the origins of the nuclear capacity demand equation used to forecast the quantity of nuclear generating capacity demanded in various time periods. The forecast procedure was not as direct as this objective might first suggest. Instead of immediately writing

$$Q_n = f(\alpha, \beta, \gamma, \delta, \ldots) \tag{1}$$

and then going about the business of estimating coefficients, we wrote

$$Q_n = \frac{Q_n}{Q} Q \tag{2}$$

where

$$\frac{Q_n}{Q} = g(A, B, C), \tag{3}$$

and

$$Q = h(X, Y, Z). \tag{4}$$

Equations 3 and 4 were then estimated separately. Projections for independent variables X, Y, and Z were obtained and forecasts of Q made using the $h(\ )$ function. This calculation made it possible to treat Q as a parameter whose value at any point in time is determined outside the model. Thus Equation 1 becomes

$$Q_n = f(A, B, C) = Q \cdot g(A, B, C). \tag{5}$$

Important preliminary matters are the spatial and temporal market boundaries for demand. In preparing the two capacity equations we used the EEI's regional partition (9 power regions) and a fifteen-year time horizon broken up into three five-year periods (1958–1962,

1963–1967, and 1968–1972). Thus we had 27 space time "cells," each of which was to be the source of an observation vector. In practice the total capacity variable had 27 observations but the nuclear share variable had zero observations in 14 of the cells. The 14 zero-share observations were ignored. Hence the total capacity regression "explains" 27 observations while the nuclear share regression "explains" but 13.

Forecasts were made from the fitted equations for every power region for each of the following five-year periods: 1985–1989, 1990–1994, 1995–1999, and 2000–2004. Thus there are three different sets of problems to be discussed. The first two relate to the estimation of 3 and 4, the last to the projection of X, Y, and Z, the independent variables in the total capacity equation. Each of these will be discussed, seriatim, below.

The Nuclear Share Equation

The hypothesized nuclear share relationship was

$$\frac{Q_n}{Q} = \phi(\text{PNF, PFF, PNK, PFK, S}), \tag{6}$$

where

PNF = nuclear fuel price,

PFF = fossil fuel price,

PNK = nuclear capital price,

PFK = fossil capital price,

S = average size (arithmetic mean) of units ordered.

The price variables are, of course, natural inclusions and their presence needs no explanation. The "size" variable is the mean capacity size of the units demanded in a particular cell (that is, in a particular region during a particular period). It is included in the equation to capture the effects of size of plant being purchased on the operating characteristics of the buyers' system — where the characteristics differ between the nuclear and fossil technologies. The buyer is assumed to have a particular distribution of plant sizes in mind, and will add a "large" plant when there is "room" in that distribution; the large plant is an inflexible unit, however, that requires more back-up capacity in case of downtime, and that is potentially less productive as peaking capacity in its later years of operation. In particular, the newer technology and poorer stop-start operating characteristics of nuclear plants weigh against the choice of that technology for small plants (*given* price differences).

Table C.1 Nuclear Share Equation Observations

(1)	PR (2)	Fossil Additions to Steam-Electric Capacity (3)	Additions to Nuclear Capacity (4)	Total (5)	$\frac{NK}{P_N}$ (6)	$\frac{FK}{P_F}$ (7)	$\frac{NFA}{F_{N_A}}$ (8)	$\frac{NFB}{F_{N_B}}$ (9)	$\frac{FF}{F_F}$ (10)	$\frac{S_1}{S}$ (11)	$\frac{S_2}{PCT}$ (12)
58-60	1	1,213^{10}	175^{1}	1,388^{11}	217	201	4.13	4.13	3.23	126	0
	2	7,101	-	7,101	-	-	-	-	-	-	-
	3	10,569^{57}	200^{1}	10,769^{58}	255	163	4.45	4.45	2.33	186	0
	4	2,805	-	2,805	-	-	-	-	-	-	-
	5	7,676	-	7,676	-	-	--	-	-	-	-
	6	4,543	-	4,543	-	-	-	--	-	-	-
	7	6,247	-	6,247	-	-	--	-	-	-	-
	8	1,472	-	1,472	-	-	-	-	-	-	-
	9	3,675	-	3,675	-	-	-	-	--	-	-
63-67	1	1,732	-	1,732	-	-	--	-	-	-	-
	2	4,744^{13}	295^{2}	5,039^{15}	387	147	3.94	1.60	2.45	336	0.330
	3	7,663^{25}	81^{2}	7,744^{27}	432	152	10.23	10.23	2.96	287	0.589
	4	3,527^{23}	81^{2}	3,608^{25}	383	198	10.76	10.76	3.18	144	0
	5	11,410	-	11,410	-	-	-	-	-	-	-
	6	5,271	-	5,271	-	-	-	-	-	-	-
	7	9,294	-	9,294	-	-	-	-	-	-	-
	8	2,072	-	2,072	-	-	-	-	-	-	-
	9	6,629^{21}	1,289^{3}	7,918^{24}	329	107	2.76	1.70	3.00	330	0.188
68-72	1	4,721^{10}	2,950^{5}	7,671^{15}	143	146	3,63	1.52	3.07	511	0.607
	2	12,409^{23}	6,923^{9}	19,332^{32}	144	134	1.79	1.51	2.42	604	0.870
	3	17,343^{50}	7,197^{11}	24,540^{61}	129	150	1.40	1.51	1.68	402	0.603
	4	9,434^{30}	1,472^{3}	10,906^{33}	168	117	3.46	1.52	2.33	330	0.212
	5	17,633^{38}	6,246^{8}	23,879^{46}	131	128	2.41	1.51	1.80	519	0.693
	6	15,805^{43}	4,260^{4}	20,065^{47}	108	126	0.46	0.46	1.68	427	0.561
	7	17,124^{49}	-	17,124	-	-	-	-	-	-	-
	8	8,028^{20}	-	8,028	-	-	--	-	-	-	-
	9	6,826^{17}	1,522^{2}	8,348^{19}	159	106	1.15	1.51	2.96	439	0.480

Notes: Small numbers superscripted refer to the (adjusted) number of units added.

The specific form of Equation 6 which actually fitted in the model was

$$\log\left(\frac{Q_n}{Q}\right) = -4.375 + 0.418 \log S - 0.859 \log\left(\frac{\text{PNF}}{\text{PFF}}\right) - 1.030 \log\left(\frac{\text{PNK}}{\text{PFK}}\right). \quad (7)$$

Observe that the coefficients of the price ratio terms are negative in sign as they should be and that the size coefficient is positive but less

Table C.2 Nuclear Plant Data

Manufacturing	PR	Thermal MW	Electrical MW	Year Critical	Year Commercially Operative	Investment $10^6	Nuclear Fuel Costs Mills/kWh
(1)	(2)	(3)	(4)	(5)	(6)	(7)	(8)
W	1	482	175	60	61	38	8.0
W	1	1473	462		68	85	
GE	1	1727	549		69	65	
GE	1	1682	625		71	83	
GE	1	1593	514		71	88	
CE	1	2440	800		72	100	
W	2	890	231	55	57	72	64.4[1]
BW	2	585	265	62	63	90	10.0
GA	2	115	40	66	67	28	
GE	2	1600	515	67	68	68	4.05
GE	2	1538	500		68	90	6.67
W	2	1300	420		69	75	
BW	2	2758	873		69	108	4.0
W	2	3250	999		71	139	
GA	2	3294	1065		71	138	
GE	2	2361	755		71	100	
BW	2	3025	965		71	159	
BW	2	2452	831		71	116	
BW	2	2452	800		73	100	
W	2	3250	993		73	121	
GE	2	2436	829		73	130	
GE	2	1593	540		73	100	
GA	2	3294	1065		73	125	
W	2	3253[2]					
GE	3	626	200	59	60	51	9.0
AI	3	45	11	63	63	8	
GE	3	157	70	62	63	27	17.11
AC	3	165	50	66	68	18	

than 1.0. This latter result, it should be noted, is consistent with the comparative performance hypothesis mentioned above. The data for fitting the equation are discussed in turn, from left to right in the final equation.

Table C.2 *Continued*

(1)	(2)	(3)	(4)	(5)	(6)	(7)	(8)
GE	3	2255	715		69	76	4.0
CE	3	2212	700		70	102	
GE	3	2255	715		70	79	4.0
W	3	1395	455		70	60	
GE	3	2255	715		70	80	
W	3	1395	455		71	57	
GE	3	2255	715		71	80	
W	3	3250	1050		72	164	
W	3	1396	527		72	61	
W	3	3250	1100		72	150[3]	
W	3	3250	1100		73	150[3]	
W	3	3250	1050		73	153	
GE	3	1593	515		73	91	
AC	4	73	22	62	64	10	19.13
AC	4	201	59	64	67	21	17.11
GE	4	1469	472		70	80	
CE	4	1420	450		71	70	
W	4	1650	550		72	98	
GE	4	2331	778		73	73	
W	4	1650	550		74	85	
W	5	2200	722		70	100	
W	5	2095	663		70	96	
W	5	2200	722		71	100	
BW	5	2452	874		71	86	
W	5	2441	783		71	130	
BW	5	2452	874		72	86	
W	5	2441	783		72	108	
BW	5	2452	825		72	110	
BW	5	2452	874		73	92	
CE	5	2450	848		73	118	

The Capacity Variables

The nuclear capacity series was built from basic and scattered sources. A history was made of the operating capacities of all nuclear reactors exclusively engaged in public power generation, whether placed in operation or scheduled to be placed in operation in the 1958–1972

Table C.2 *Continued*

(1)	(2)	(3)	(4)	(5)	(6)	(7)	(8)
W	5		800		74	108	
CE	5	2450	848		75	105	
GE	6	3293	1065		70	117[4]	
GE	6	3293	1065		70	118[4]	
GE	6	3293	1065		72	115	
GE	6	3293	1065		72	115[5]	
BW	7	2452	800		73	140	
GA	8	835	330		73	69	
GE	9	165	69	63	63	22	8.0
GE	9	4000	790	63	66	315	
W	9	1347	430	66	67	87	6.39
W	9	3250	1060		71	150	
W	9	1473	462		72	92	4.7
BW	9	2452	800		73	134	
CE	0	18	17	63	65	12[6]	24.9

Notes

1 Shippingport plant excluded from all later calculations.

2 Deleted for lack of informat.

3 $300 was for both plants, so it was divided evenly between them.

4 235 was for both plants. In the absence of any further information the cost was divided up equally between the 2 plants.

5 All these are guesstimates.

6 Puerto Rico; excluded from all further calculations.

period. The sources were the various nuclear and electrical technical magazines (for example, *Nucleonics*) where reactor news is carried. In addition to providing a master listing of capacities, starting dates, estimates of fuel, and capital costs were found in a number of cases. All this information is shown in undigested form in Table C.2. Table C.3, on the other hand, is exclusively concerned with nuclear capacity and was obtained by aggregating the individual capacities given in Table C.2 by region and by period. It is the immediate source for column 4 of Table C.1.

The total (steam-electric) generating capacity series given in Table C.1 (column 5) is, of course, equal to the sum of the nuclear and fossil

Table C.3 Total Additions to United States Nuclear Capacity by Region and by Period (see footnote 3 of this appendix).

	Period											
PR (1)	1957 and earlier (2)		1958-1962 (3)		1963-1967 (4)		1968-1972 (5)		1973 and beyond (6)		Total (7)	
0	-	-	-	-	-[2]	-	-	-	-	-	-[2]	-
1	-	-	175	1	-	-	2,950	5	-	-	3,125	6
2	-[1]	-	-	-	295	2	6,923	9	4,227	5	11,445[1]	16
3	-	-	200	1	81	2	7,197	11	2,665	3	10,143	17
4	-	-	-	-	81	2	1,472	3	1,328	2	2,881	7
5	-	-	-	-	-	-	6,246	8	3,370	4	9,616	12
6	-	-	-	-	-	-	4,260	4	-	-	4,260	4
7	-	-	-	-	-	-	-	-	800	1	800	1
8	-	-	-	-	-	-	-	-	330	1	330	1
9	-	-	-	-	1,289	3	1,522	2	800	1	3,611	6
Total	-[1]	-	375	2	1,746[2]	9	30,570	42	13,520	17	46,211[1,2]	7

Notes

1 Shippingport plant excluded (231 MW_e).

2 Puerto Rican Plant excluded (17 MW_e).

3 Compiled by Stull, 7/8/68. Ultimate source--master list of additions to capacity.

components. The nuclear series has been discussed in the last paragraph; the fossil series can now be reviewed.

The basic source for the fossil series was the National Coal Association's annual publication *Steam-Electric Plant Factors.*[1] Table 4 in each of these publications provides a list of scheduled capacity additions, by year of completion and by state, to existing steam-electric generating capacity. These additions are either in the form of new plants or new units in existing plants. The data provided in these tables derived ultimately from the Federal Power Commission, which requires the major electrical utility systems to report all scheduled additions to existing capacity (Form 12E). These systems account for "about 98 per cent of total (U.S.) utility requirements."

[1] Henceforth to be called *Factors.*

The time horizon in any single *Factors* issue was usually four years. Thus the 1963 issue would list capacity additions for the years 1964–1967. As might be expected of any "forecast additions" series, the lists in different issues of *Factors* very often turned out to be inconsistent with one another. For example, the 1962 list would show that a 150 MW_e addition to Plant *X* in State *Y* was scheduled to be completed in 1964. The 1963 *Factors*, on the other hand, might indicate that the completion year was 1965 instead of 1964 or that the capacity to be added was 175 MW_e instead of 150. In the light of such inconsistencies it was decided that only the latest series for each year would be used. This meant that for the years prior to 1968 the appropriate series was that provided in the *Factors* immediately preceding the year in question. Thus the 1964 series which was actually used was the one found in the 1963 *Factors*. For 1968 and later years, however, the series provided in the 1966 *Factors* had to suffice since they were the latest available.

Having made this decision, it first appeared that the resulting series could be aggregated into five-year period totals for the nine power regions. This was not the case; three further complications arose.

The first was relatively minor, but a few preliminary explanations are required before disposing of it. It should be realized at the outset that the purpose for constructing the series here is to build a forecast model of choice of one energy producing technology rather than another. Therefore, it is appropriate to focus only on those investment decisions which actually induce this choice. Some of the capacity additions listed in *Factors* represent so-called "stretching" operations. Such operations involve the making of minor technical and structural changes in an existing generating plant, the usual result being only marginal improvements in its capabilities. Such changes, it would seem, are not the consequences of decisions on technology. In light of this the stretching operations had to be deleted. The following procedure was adopted:

a. All capacity additions less than 50 MW_e to *existing* plants were regarded as "stretching" operations. All additions of size 50 MW_e and greater to existing plants and all new plant capacity were assumed to be the consequences of decisions directly concerned with the nuclear-fossil issue and thus to be included.

b. Capacity increments designated as "stretching" were then excluded from consideration unless the parent plant was constructed after 1957, in which case they were allocated to the completion year and power region of the parent plant.

Thus total capacity here is the sum of three components: total new plant capacity, additions to existing plants of size 50 MW_e and larger, and "stretching" capacity added in later periods to plants newly built

Table C.4a Additions to Fossil Steam-Electric Capacity, Final Figures 1958–1962 Period

PR	1958	1959	1960	1961	1962	Totals
1	325.0^{3}	225.0^{2}	375.0^{3}	287.5^{2}	-	$1,212.5^{10}$
2	$1,642.3^{10}$	$1,348.3^{7}$	$1,958.3^{8}$	$1,007.0^{3}$	$1,145.0^{5}$	$7,100.9^{33}$
3	$2,361.6^{15}$	$3,129.0^{18}$	$1,350.0^{7}$	$1,680.0^{8}$	$2,048.8^{9}$	$10,569.4^{57}$
4	441.3^{5}	656.3^{5}	656.6^{7}	790.0^{7}	260.6^{3}	$2,804.8^{27}$
5	$1,997.0^{16}$	$1,327.5^{9}$	$1,857.0^{9}$	962.0^{6}	$1,532.6^{7}$	$7,676.1^{47}$
6	540.0^{3}	$2,069.0^{11}$	673.7^{5}	750.0^{2}	509.9^{3}	$4,542.6^{24}$
7	$1,564.2^{13}$	$1,467.2^{11}$	$1,207.7^{9}$	$1,197.6^{6}$	810.1^{5}	$6,246.8^{44}$
8	219.0^{3}	154.0^{2}	439.0^{5}	229.2^{2}	431.0^{8}	$1,472.2^{20}$
9	892.5^{7}	604.2^{4}	436.5^{2}	795.2^{4}	946.5^{4}	$3,674.9^{21}$
Totals	$9,982.9^{75}$	$10,980.5^{69}$	$8,953.8^{55}$	$7,698.5^{40}$	$7,684.5^{44}$	$45,300.2^{283}$

Table C.4b Additions to Fossil Steam-Electric Capacity, Final Figures 1963–1967 Period

PR	1963	1964	1965	1966	1967	Totals
1	375.0^{2}	461.0^{2}	493.8^{2}	22.0^{1}	380.0^{1}	$1,731.8^{8}$
2	$1,237.5^{4}$	695.2^{3}	$1,150.0^{2}$	179.5^{1}	$1,482.1^{3}$	$4,744.3^{13}$
3	672.6^{2}	$1,119.4^{4}$	$1,354.3^{4}$	756.4^{4}	$3,760.0^{11}$	$7,662.7^{25}$
4	310.0^{4}	778.3^{6}	226.2^{2}	712.0^{5}	$1,500.3^{6}$	$3,526.8^{23}$
5	$1,185.0^{5}$	$2,279.9^{9}$	$2,832.0^{9}$	$2,680.9^{10}$	$2,532.0^{6}$	$11,409.8^{39}$
6	$1,721.0^{4}$	151.0^{1}	$1,499.2^{4}$	449.3^{2}	$1,450.0^{2}$	$5,270.5^{13}$
7	$1,262.0^{6}$	723.4^{5}	$1,490.1^{6}$	$2,870.7^{13}$	$2,948.0^{6}$	$9,294.2^{36}$
8	559.0^{4}	714.7^{4}	510.0^{4}	-	288.4^{2}	$2,072.1^{14}$
9	$1,150.0^{4}$	$1,430.0^{6}$	826.0^{4}	$1,280.0^{3}$	$1,943.0^{4}$	$6,629.0^{21}$
Totals	$8,472.1^{35}$	$8,352.9^{40}$	$10,381.6^{37}$	$8,950.8^{39}$	$16,183.8^{41}$	$52,341.2^{192}$

Table C.5 Additions to Fossil Steam-Electric Capacity: 1968–1972 by Regions

Figure A: Uninflated (Adjusted for stretching and double-counting only)

PR	1968	1969	1970	1971	1972
1	1,235.1^{3}	595.0^{1}	-	-	-
2	3,122.5^{5}	1,297.8^{3}	1,950.3^{3}	900.0^{1}	1,305.0^{2}
3	4,282.1^{12}	1,331.0^{5}	4,297.0^{10}	1,370.0^{2}	-
4	1,340.0^{4}	1,228.8^{4}	1,151.0^{5}	1,020.0^{2}	-
5	2,625.4^{6}	3,140.0^{6}	3,810.0^{8}	1,626.0^{3}	434.0^{1}
6	365.2^{2}	3,108.7^{7}	-	420.0^{1}	-
7	3,064.0^{11}	2,914.1^{7}	4,237.0^{11}	1,385.0^{4}	235.0^{1}
8	811.6^{5}	755.0^{1}	2,460.0^{4}	755.0^{1}	-
9	975.0^{3}	-	-	-	-

Figure B: Final Figures

PR	1968	1969	1970	1971	1972	Total
1	1,266.0^{3}	622.4^{1}	944.2^{2}	944.2^{2}	944.2^{2}	4,721.0^{10}
2	3,534.7^{6}	1,428.9^{3}	2,481.8$^{4.5}$	2,481.8$^{4.5}$	2,481.8$^{4.5}$	1[illegible],409.0$^{22.5}$
3	4,958.7^{14}	1,609.2^{6}	4,207.0^{10}	3,284.0^{10}	3,284.0^{10}	17,342.9^{50}
4	1,626.8^{5}	2,146.7^{7}	1,886.8^{6}	1,886.8^{6}	1,886.8^{6}	9,433.9^{30}
5	2,898.4^{7}	4,012.9^{8}	3,810.0^{8}	3,455.7$^{7.5}$	3,455.7$^{7.5}$	17,632.7^{38}
6	651.5^{4}	5,670.3^{13}	3,160.9$^{8.5}$	3,160.9$^{8.5}$	3,160.9$^{8.5}$	15,804.5$^{42.5}$
7	3,171.2^{11}	3,272.5^{8}	4,237.0^{11}	3,221.8$^{9.5}$	3,221.8$^{9.5}$	17,124.3^{49}
8	992.6^{6}	1,791.6^{2}	2,460.0^{4}	1,392.1^{4}	1,392.1^{4}	8,028.4^{20}
9	979.9^{3}	1,461.5$^{3.5}$	1,461.5$^{3.5}$	1,461.5$^{3.5}$	1,461.5$^{3.5}$	6,825.9^{17}

in this period. Parenthetically, it might be noted that the last category turned out to be of little empirical importance.

The second complication followed from concern with the adequacy of the National Coal Association data. It turned out, as mentioned above, that there were differences between two consecutive issues of

Factors on the completion date for a particular project. In this case the first entry was redundant and had to be excluded from the cell total. Because the potential distortions involved were in some cases very large, all capacity series were closely scrutinized to be sure all such double-counting had been eliminated. The resulting steam-electric capacity figures for the two earlier periods are given in Tables C.4a and b, and are, of course, fully adjusted to take into account the "stretching" and "double-counting" problems. The first 18 regional totals in column 2 of Table C.1 are immediately derived from this table.

The final problem is the quality of the 1966 *Factors* series for 1968–1972. Given the numerous changes in the earlier series, it was feared that the 1968–1972 "planned additions" would not be realized. They were, as a consequence, adjusted to improve this reliability. The general method was to assay the quality of past predictions and then use these results to inflate the 1966 numbers.

Consider as a specific example the testing and "correction" of the 1968 series given in Table C.5, Figure A. To begin, consider Table C.6. This table clearly reveals that the National Coal Association's two-year predictions from 1958 to 1965 generally underestimated what in fact took place and often by considerable margins (the "actualities" here being the one-year predictions mentioned earlier). Since the deviations were systematic and not too large, some escalation of the 1968 series was not out of order and would on balance lead to a better forecast result. Further, since the quality of the forecasting performances varied considerably between power regions (see last row, Table C.6), we thought that the regions should be treated separately. The inflater actually used was the ratio of actual total capacity added from 1960 through 1967 to predicted total capacity added for the same period. Thus, in Table C.6, predicted added capacity for Power Region 1 was 2,406 MW_e while actual capacity added was 2,466 MW_e (the ratio being 1.025). The adjusted capacity figure ("adjusted" in the sense that all stretching and double-counting entries were excluded) given in Table C.5, Figure A (1,235.1) was then multiplied by 1.025 to get the final 1968 Power Region 1 estimate (1,266.0). This process was then repeated for the other eight regions.

The inflation of the 1969 series proceeded in exactly the same way with but one exception (see Tables C.5 and C.7). The 1966 *Factors* predicted that zero additional capacity was going to be added in Power Region 9 in 1969. This appeared to be an unreasonable result but it obviously could not be cured by the method previously employed. Hence an alternative procedure was devised: the 1967–1968 average was used as the 1969 entry.

Table C.6 Predicted and Actual Additions to Fossil Capacity

		Power Regions								
		1	2	3	4	5	6	7	8	9
1	1958 version of 1960	388	1,365	885	612	1,778	1,141	1,250	408	667
	actual 1960	375	1,975	1,405	657	1,857	1,141	1,208	483	654
2	1959 version of 1961	275	1,257	1,762	779	1,199	250	1,418	229	578
	actual 1961	288	1,314	1,804	823	1,199	750	1,198	229	795
3	1960 version of 1962	-	1,065	1,869	260	1,183	1,933	676	260	1,237
	actual 1962	-	1,477	2,065	339	1,533	1,925	810	475	947
4	1961 version of 1963	375	1,238	696	337	995	156	1,181	500	1,120
	actual 1963	375	1,238	695	337	1,185	2,221	1,544	559	1,150
5	1962 version of 1964	472	682	1,434	695	2,114	751	1,058	601	1,474
	actual 1964	488	695	1,216	778	2,280	1,251	1,155	715	1,545
6	1963 version of 1965	516	1,150	1,302	390	2,832	1,199	1,386	510	810
	actual 1965	516	1,150	1,387	478	2,936	1,499	1,801	510	826
7	1964 version of 1966	-	180	1,509	1,168	2,951	1,399	2,622	206	1,280
	actual 1966	44	180	1,527	1,201	3,209	1,899	2,871	206	1,280
8	1965 version of 1967	380	1,517	2,524	859	2,030	-	3,498	137	1,930
	actual 1967	380	1,542	3,776	1,576	2,454	1,495	2,962	311	1,943
		2,466	9,571	13,875	6,189	16,653	12,181	13,549	3,488	9,140
		2,406	8,454	11,981	5,100	15,082	6,829	13,089	2,851	9,096
		1.025	1.132	1.158	1.214	1.104	1.784	1.035	1.223	1.005

Table C.7 Predicted and Actual Additions to Fossil Capacity

		Power Regions								
		1	2	3	4	5	6	7	8	9
1	1957 version of 1960	388	1,945	1,276	304	1,422	1,025	1,075	214	647
	actual 1960	375	1,975	1,405	657	1,857	1,141	1,208	483	654
2	1958 version of 1961	275	1,240	1,631	801	908	250	1,248	163	578
	actual 1961	288	1,314	1,804	873	1,199	750	1,198	229	795
3	1959 version of 1962	375	1,154	1,848	135	960	1,283	676	110	937
	actual 1962	0	1,477	2,065	339	1,533	1,925	810	475	947
4	1960 version of 1963	375	1,259	917	66	871	1,306	1,379	150	660
	actual 1963	375	1,238	695	337	1,185	2,221	1,544	559	1,150
5	1961 version of 1964	445	283	970	550	1,687	600	993	300	1,185
	actual 1964	488	695	1,216	778	2,280	1,251	1,155	715	1,545
6	1962 version of 1965	494	1,150	1,302	200	2,832	1,199	1,776	260	490
	actual 1965	516	1,150	1,387	478	2,936	1,499	1,801	510	826
7	1973 version of 1966	0	180	1,363	647	2,294	449	1,924	173	1,280
	actual 1966	44	180	1,527	1,201	3,209	1,899	2,870	206	1,280
8	1964 version of 1967	380	1,482	2,170	839	2,051	566	2,998	100	1,930
	actual 1967	380	1,542	3,776	1,576	2,454	1,495	2,962	311	1,943
		$\frac{2,466^*}{2,732}$	$\frac{9,571}{8,693}$	$\frac{13,875}{11,477}$	$\frac{6,189}{3,542}$	$\frac{16,652}{13,025}$	$\frac{12,181}{6,678}$	$\frac{13,548}{12,069}$	$\frac{3,488}{1,470}$	$\frac{9,140}{7,707}$
		.903	1.101	1.209	1.747	1.278	1.824	1.123	2.373	1.186

* $\frac{2,466}{2,357}$, 1.046. The third observation in Column 1 was excluded as aberrant.

For 1970 it was no longer reasonable to continue to inflate the 1966 *Factors* predictions as we had done for 1968 and 1969. In the first place the *Factors* prediction record over a four-year horizon (see Table C.8) appeared to be too inaccurate and erratic to be used safely. For example, there were many zero predictions for regions which later turned out to have substantial investments in new capacity (see last row, Table C.8). Hence, continuation of the same adjustment method resulted in gross adjustments in the *Factors* figures whereas earlier the changes made were, on the whole, marginal. Moreover, no matter what was done with the 1970 figures the method could not be applied to their 1971 and 1972 counterparts simply because the earlier editions of *Factors* (before 1965) did not provide construction data five or six years in advance.

A new approach had to be devised for the future years in any case. A stopgap procedure, already used in the case of Power Region 9 in 1969, was adopted for the years 1970, 1971, and 1972. The first step was to calculate the 1968–1969 average for all regions except Region 9 where the 1967–1968 average was used. This average was then used as the 1970, 1971, and 1972 entry *unless* the predicted figure was greater (in which case the predicted figure was used). One can verify this application by comparing Figure A and Figure B of Table C.5.

The final step was, as before, the addition of annual totals to attain regional five-year totals. This is done in the last column of Figure B, Table C.5, the immediate source for the final nine entries in column 3, Table C.1.

The Size of the Typical New Plant

Two alternative *size* variables were tried in the course of fitting the regression equations. The first was simply the mean unit size added in a cell (Table C.1, column 11); the second was the percentage of total capacity added in a cell by units 500 MW_e or greater in size (Table C.1, column 12). The first of these was the one finally settled upon and is the one which will be discussed below.

For the first two periods the size variable presented no new problems since it was simply the total capacity added in a cell (fossil and nuclear) divided by the total number of units added. These calculations were, of course, made after all double-counting and stretching entries were expunged. Tables C.4a and C.4b show the build-up of the fossil total unit entries in column 3 of Table C.1 for these two periods. Their nuclear counterparts in column 4 of the same table are derived for all three periods from Table C.3 which is in turn merely an aggregation of the contents of the master reactor list (Table C.2).

Table C.8 Predicted and Actual Additions to Fossil Capacity

		Power Regions								
		1	2	3	4	5	6	7	8	9
1	1956 version of 1960	0	1,543	845	229	375	525	1,095	0	300
	actual 1960	375	1,975	1,405	657	1,857	1,141	1,208	483	654
2	1957 version of 1961	0	860	820	533	750	225	1,233	0	300
	actual 1961	288	1,314	1,804	823	1,199	750	1,198	229	795
3	1958 version of 1962	375	751	996	60	200	950	595	0	200
	actual 1962	0	1,477	2,065	339	1,533	1,925	810	475	947
4	1959 version of 1963	0	513	652	144	441	500	587	0	200
	actual 1963	375	1,238	695	337	1,185	2,221	1,544	559	1,150
5	1960 version of 1964	0	296	580	314	0	0	535	150	640
	actual 1964	488	695	1,216	778	2,280	1,251	1,155	715	1,545
6	1961 version of 1965	0	400	580	0	1,341	900	0	0	365
	actual 1965	516	1,150	1,387	478	2,936	1,499	1,801	510	826
7	1962 version of 1966	0	200	673	675	2,486	479	324	150	1,065
	actual 1966	44	180	1,527	1,201	3,209	1,899	2,871	206	1,280
8	1963 version of 1967	0	1,578	1,165	498	1,284	500	518	200	1,480
	actual 1967	380	1,542	3,776	1,576	2,454	1,495	2,962	311	1,943
		-	9,571/6,141	13,875/6,311	6,189/2,453	16,653/6,877	-	13,549/4,887	3,488/500	-
			1.559	2.199	2.523	2.422		2.772	6.976	

The 1968–1972 fossil entries, however, required some further work. It was not possible to use an adjusted enumeration out of the 1966 *Factors* of units added (as presented in Table C.5a) because the numerator of the fraction had been inflated in the process of developing a total capacity series. Hence, we decided to treat the fossil unit figures in the same way as we treated the fossil capacity figures.

Thus the final 1968 and 1969 plants (except for 1969, Region 9) equal the predicted count times the regional adjustment factor used to inflate the total fossil capacity predictions (the result being then rounded to the nearest unit). Similarly, the 1970–1972 final figures equal the 1968–1969 average (except for Region 9) unless the predicted count is higher in which case it is used. For Power Region 9 the 1967–1968 average is used for all years from 1969–1972. The results of these transformations can be seen by comparing Table C.5b to C.5a. Table C.5b is thus the source for the 1968–1972 fossil unit entries given in column 3, Table C.1.

In summary, then, the 13 size observations in column 11, Table C.1 are simply the ratios of the 13 sets of capacity and plant number estimates given in column 5 of the same table. These latter numbers are equal to the sums of their counterparts in columns 3 and 4.

Fossil Fuel and Capital Prices

The general approach here was to associate a "representative" fossil plant with each nuclear plant ordered. The fuel and capital prices for these selected fossil plants were then averaged to get the cell figures given in columns 7 and 10 of Table C.1. The discussion which follows will deal mainly with the selection of the representative fossil plant.

First of all, a listing was made for each nuclear plant of all fossil plants in the same power region which were within a (80–120 per cent) size range centered on the size of the nuclear plant in question. Occasionally, a different selection principle (for example, 70–130 per cent range) was used in order to get a reasonable number of observations. In such atypical cases a special note was made on the worksheet. The lists were compiled from the National Coal Association's annual *Factors* series. If the first year of commercial operation of the nuclear plant was 1957 or earlier, then the 1957 *Factors* was used. If it was between 1958 and 1962 (inclusive) the 1962 *Factors* was used. Finally, if the first year of commercial operation of the nuclear plant was 1963 or later, then the 1966 *Factors* (the most recent available) was used.

The next step was to obtain fuel and capital cost data and the initial year of operation for each such selected fossil plant. The source here was an annual FPC publication, *Steam-Electric Plant Construction*

Table C.9 Fossil Capital and Fossil Fuel Cost Estimates

PR (1)	Nuclear Plant, Name, State (2)	Size, MW_e (3)	Year (4)	Fuel Cost Estimates: Low Cost of 5 Most Recent (5)	Capital Cost Estimates (dollars per kwe of installed capacity) (6)	Low Fuel Cost Plant of 5 Most Recent: Name, State (7)	Size MW_e (8)	Initial Year of Operation (9)
1	Yankee, Mass.	175	1961	3.23	$201	Norwalk H., Conn.	150.0	1960
1	Connecticut Yankee, Conn.	462	1968	3.07	146	Brayton Pt., Mass.	482.0	1963
1	Millstone, Conn.	549	1969	3.07	146	Brayton Pt., Mass.	482.0	1963
1	Vermont Yankee, Vt.	514	1971	3.07	146	Brayton Pt., Mass.	482.0	1963
1	Pilgrim, Mass.	625[1]	1971	3.07	146	Brayton Pt., Mass.	482.0	1963
1	Maine Yankee, Me.	800[1]	1972	3.07	146	Brayton Pt., Mass.	482.0	1963
2	Shippingport, #1,#2, Penn.	231	1957	1.90	$136	Shawville, Penn.	265.0	1954
2	Peach Bottom #1, Penn.	40[2]	1967	2.01	166	Milesburg, Penn.	46.0	1950
2	Nine Mile Pt., N. Y.	500	1968	1.91	161	Elrama, Penn.	425 (532.8)[8]	1952
2	Oyster Creek, N. J.	515	1968	1.83	135	Mitchell, Penn.	448.7	1948
2	Shoreham, N. Y.	540	1973	1.75	137	Shawville, Penn.	640.0	1954
2	Robert E. Ginna, N. Y.	420	1969	2.68	162	Portland, Penn.	394.5 (383.0)[3]	1958
2	Easton, N. Y.	755	1971	1.75	137	Shawville, Penn.	640.0	1954
2	Keyport (Jersey Central P & L), Penn.	800	1973	1.75	137	Shawville, Penn.	640.0	1954

(1)	(2)	(3)	(4)	(5)	(6)	(7)	(8)	(9)
2	Milliken (Bell), N. Y.	829	1973	2.41	104	Brunner Is., Penn.	768.3	1961
2	Goldsboro, Penn.	831	1971	2.41	104	Brunner Is., Penn.	768.3	1961
2	Indian Pt. #2, N. Y.	873	1969	2.41	104	Brunner Is., Penn.	768.3	1961
2	Indian Pt. #3, N. Y.	965	1971	2.75	142	Huntley, N. Y.	828.0	1916/ 1942
2	Burlington #2, N. J.	993	1973	2.75	142	Huntley, N. Y.	828.0	1916/ 1942
2	Burlington #1, N. J.	999	1971	2.75	142	Huntley, N. Y.	828.0	1916/ 1942
2	Peach Bottom #2, Penn.	1065[4]	1971	2.75	142	Huntley, N. Y.	828.0	1916/ 1942
2	Peach Bottom #3, Penn.	1065[4]	1973	2.75	142	Huntley, N. Y.	828.0	1916/ 1942
2	Indian Pt. #1, N. Y.	265	1963	2.52	144	Milliken, N. Y.	270.0	1955
3	Dresden #1, Ill.	200	1960	2.33	$163	Vermillion, Ill.	182.3	1955
3	Piqua, Oh.	11[5]	1963	1.77	194[9]	Jasper, Ind.	9.5	unknown
3	Big Rock Pt., Mich,	70[6]	1963	3.15	145	Stoneman, Wyo.	51.8	1950
3	Lacrosse, Wisc.	50[7]	1968	2.14	292	Culley, Ind.	40.0	1955
3	Point Beach #1, Wisc.	455	1970	1.66	157	Breed, Ind.	450.0	1960
3	Point Beach #2, Wisc.	455	1971	1.66	157	Breed, Ind.	450.0	1960
3	Bailly, Ind.	515	1973	1.66	157	Breed, Ind.	450.0	1960
3	Kewaunee, Wisc.	527	1972	1.66	157	Breed, Ind.	450.0	1960

Table C.9 *Continued*

(1)	(2)	(3)	(4)	(5)	(6)	(7)	(8)	(9)
3	Dresden #2, Ill.	715	1969	1.68	154	Sammis, Oh.	740.0	1959
3	Dresden #3, Ill.	715	1970	1.68	154	Sammis, Oh.	740.0	1959
3	Palisades, Mich.	700	1970	1.68	154	Sammis, Oh.	740.0	1959
3	Quad Cities #1, Ill.	715	1970	1.68	154	Sammis, Oh.	740.0	1959
3	Quad Cities #2, Ill.	715	1971	1.68	154	Sammis, Oh.	740.0	1959
3	Zion #1, Ill.	1050	1972	1.65	116	Muskingam, Oh.	876.0	1953
3	Bridgeman #1, Mich.	1100	1972	1.71	152	Joppa, Ill.	1100.3	1953
3	Zion #2, Ill.	1050	1973	1.65	116	Muskingam, Oh.	876.0	1953
3	Bridgeman #2, Mich.	1100	1973	1.71	152	Joppa, Ill.	1100.3	1953
4	Elk River, Minn.	22[10]	1964	3.24	$259	Ben French, S. D.	22.0	1960
4	Pathfinder, S. D.	59[11]	1967	3.16	175	Bridgeport, Ill.	71.0	1953
4	Monticello, Minn.	472	1970	2.17	114	Montrose, Mo.	563.1	1958
4	Ft. Calhoun, Neb.	450	1971	2.70	125	Black Dog, Minn.	486.7	1952
4	Prairie Islands #1, Minn.	550	1972	2.17	114	Montrose, Mo.	563.1	1958
4	Prairie Islands #2, Minn.	550	1974	2.17	114	Montrose, Mo.	563.1	1958
4	Cooper, Neb.	778[12]	1973	2.06	175	Mercmac, Mo.	923.1 (800.0)	1953
5	Robinson #2, S. C.	663	1970	1.54	$129	Clinch R., Va.	669.0	1958
5	Turkey Point #3, Fla.	722	1970	1.54	129	Clinch R., Va.	669.0	1958
5	Turkey Point #4, Fla.	722	1971	1.54	129	Clinch R., Va.	669.0	1958

(1)	(2)	(3)	(4)	(5)	(6)	(7)	(8)	(9)
5	Surry #1, Va.	783	1971	1.54	129	Clinch R., Va.	669.0	1958
5	Surry #2, Va.	783	1972	1.54	129	Clinch R., Va.	669.0	1958
5	Surry #3, Va.	800	1973	1.54	129	Clinch R., Va.	669.0	1958
5	Crystal River #3, Fla.	825	1972	1.54	129	Clinch R., Va.	669.0	1958
5	Calvert Cliffs #1, Md.	843	1973	2.46	124	Chalk Pt., Va.	727.6	1964
5	Calvert Cliffs #2, Md.	843	1975	2.46	124	Chalk Pt., Va.	727.6	1964
5	Ocones #1, S. C.	874	1971	2.46	124	Chalk Pt., Va.	727.6	1964
5	Ocones #2, S. C.	874	1972	2.46	124	Chalk Pt., Va.	727.6	1964
5	Ocones #3, S. C.	874	1973	2.46	124	Chalk Pt., Va.	727.6	1964
6	Browns Ferry #1, Ala.	1065	1970	1.68	$126	Gallatin, Tenn.	1255.2 (1050.0)[14]	1956
6	Browns Ferry #2, Ala.	1065	1972	1.68	126	Gallatin, Tenn.	1255.2 (1050.0)[14]	1956
6	Browns Ferry #3, Ala.	1065	1972	1.68	126	Gallatin, Tenn.	1255.2 (1050.0)[14]	1956
7	Ark. Power & Light, Ark.	800	1973	1.88	$ 73	Bertron, Tex.	826.2	1956
8	Fort St. Vrain, Col.	330	1973	1.97	$ 95	Cunningham, N. M.	265.4	1957

Table C.9 *Continued*

(1)	(2)	(3)	(4)	(5)	(7	(7)	(8)	(9)
9	Humboldt Bay, Cal.	69[13]	1963	3.47	$103[15]	Glenarm, Cal.	80.3	
9	N. P. R., Wash.	790	1966	2.91	102	Haynes, Cal.	920.0	1962
9	Rancho Seco, Cal.	800	1973	2.91	102	Haynes, Cal.	920.0	1962
9	San Onofre, Cal.	430	1967	3.08	116	Mandalay, Cal.	435.2	1959
9	Malibu, Cal.	462	1972	3.08	116	Mandalay, Cal.	435.2	1959
9	El Diablo, Cal.	1060	1971	2.91	102	Haynes, Cal.	920.0	1962

Notes

1 No fossil plants w/i 20% range for this power region. See power region work sheets for criteria used.

2 Used 30% range -- only 2 plants w/i this range for which data available.

3 1966 Factors size (= 394.5) differs from 1965 Construction Cost size (= 383). Cost figures for latter size.

4 No fossil w/i 20% range in the power region. See worksheets for criteria used.

5 Complete plant data unavailable for this size range. See worksheet for estimation method (fuel cost only).

6 Used 30% range -- only 6 plants within this range (out of 20) for which data was available.

7 Data available for only approximately 1/4 of plants within 20% range. Choice made from within this subset. Also, fuel cost figure is of 1961 vintage.

8 1966 Factors size (= 425) differs from 1965 Construction Cost size (= 532.8). Cost figures are for latter size plant.

9 Total value of Jasper production plant obtained from _Statistics of Electrical Utilities in the United States_, 1963. Publicly owned P.14. Capital Cost figure then obtained by division.

10 Data available for only approximately 1/3 of the plants within 20% range. Ben French is the selection from this subset, using the standard methods thereon.

11 Data available for only approximately 1/2 of plants within 20% range. Choice made from this subset.

12 Only one plant in 20% range.

13 Complete plant data unavailable for fossil plants in this size range. See Power Region 9 worksheet for estimation method for fuel cycle costs.

14 1966 _Factors_ size (= 1255.2) differs from 1965 _Construction Cost_ size (= 1050.0), cost figures being for the latter size.

15 Glenarm capital cost figure obtained by dividing the total plant cost for the Pasadena Power Department (Glenarm and Broadway plant cost) by its total installed generating capacity (= 176.3 in 1963 = 80.3 Glenarm and 96 Broadway). Plant costs for Pasadena obtained from _Statistics of Publicly Owned Electric Utilities_, 1963, page 6.

Cost and Annual Production Expenses.[2] It turned out, however, that this source, though apparently the best available, did not cover every plant listed in *Factors*. In particular, many small plants listed in *Factors* are not covered in *Construction Cost*. In such cases the unlisted fossil plant was ignored in favor of those fossil plants for which data were available, or cost estimates were concocted based on the incomplete data given in *Factors*. In either case the specific method used is explained on the worksheets.

Third, out of each set of fossil plants showing fuel cost data, five were chosen that had construction dates closest to the nuclear plants in question. Then of these five, the low (fuel) cost plant became the "representative" fossil plant. Table C.9 shows a complete listing of all nuclear plants with these fossil "best alternative plants."

Finally, a single estimate of fuel and capital costs for each cell had to be constructed. To do this we used a simple averaging process, with the generating capacities of the various nuclear units in the cell as weights (for these calculations see Table C.10; the underlying worksheets can be obtained from the author[3]).

Nuclear Fuel and Capital Prices

Experiments were tried with two alternative nuclear fuel price series. The first was another product of the initial search through various technical magazines. In column 4 of the nuclear fuel cycle cost worksheets[3] there are reproduced the total cost figures, in mills per kilowatt hour, obtained from these various sources. Capital cost estimates obtained from the same sources (column 5) are then subtracted to give the Variant A fuel price in column 7.

As is obvious from this table, this series is incomplete. Fuel price estimates could not be found for every nuclear plant. Hence the averaging process used above to get fossil prices for each cell could not be used here. Instead a different procedure was employed. First the size of the average nuclear unit in the cell was calculated, and then one plant in that cell was selected which did have fuel cost data and was closest in size to the cell average. The fuel cost figure for this particular unit was then used as the entry in the cell observation. In the case where there was but one unit in a cell with a fuel estimate, this figure became perforce the cell figure. In one instance, no cost figure whatsoever was available: Power Region 4, 1968–1972. Here we obtained the cell's nuclear size average as before, but then searched among the 1968–1972 cells of the other eight power regions for the plant closest in size to this

[2] Henceforth to be known as *Construction Cost*.

[3] Fuel cycle cost worksheets, in mimeograph form.

Table C.10 Calculation of Regional Price Sets by Period

Period	Nuclear Plant, Name and State	PR	Size, MW_e	Fossil Fuel Costs mills/kWh	Col. 1 x Col. 2	Fossil Capital Costs $/kW$_e$	Col. 1 x Col. 4	Nuclear Capital Costs $/kW$_e$	Col. 1 x Col. 6	Nuclear Fuel Costs mills/kWh A	Nuclear Fuel Costs mills/kWh B	Fossil Prices: Fuel/ Capital	Nuclear Prices: Fuel/Capital A B
			(1)	(2)	(3)	(4)	(5)	(5)	(7)	(8)	(9)	(10)	(11)
58-62	Yankee, Mass.	1	175	3.23		201		217		4.13			
	Totals		175									3.23/201	4.13/217 4.13/217
63-67													
68-72	Conn. Yankee, Conn.	1	462	3.07		146		184	85,008	2.72	1.52		
	Millstone, Conn.	1	549	3.07		146		118	64,782		1.52		
	Vt. Yankee, Vt.	1	514	3.07		146		171	87,894		1.52		
	Pilgrim, Mass.	1	625	3.07		146		133	83,125	3.63	1.52		
	Maine Yankee, Me.	1	800	3.07		146		125	100,000		1.51		
	Totals		2950 (590)						420,809			3.07/146	3.63/143 1.52/143
72-77													

Table C.10 *Continued*

			(1)	(2)	(3)	(4)	(5)	(6)	(7)	(8)	(9)	(10)	(11)
58-62													
63-67	Peach Bottom # 1, Penn.	2	40	2.01	80.40	166	6,640	700	28,000				
	Indian Point #1, N. Y.	2	265	2.52	667.80	144	38,160	340	90,100	3.94	1.60		
	Totals		305		748.20		44,800		118,100			2.45/147	3.94/387 1.60/387
68-72	Nine Mile Pt., N. Y.	2	500	1.91	955.00	161	80,500	180	90,000	3.46	1.52		
	Oyster Ck., N. J.	2	515	1.83	942.45	135	69,525	132	67,980	1.34	1.52		
	Robert Ginna, N. Y.	2	420	2.68	1,125.60	162	68,040	179	75,180		1.52		
	Easton, N. Y.	2	755	1.75	1,321.25	137	103,435	132	99,660		1.51		
	Goldsboro, Penn.	2	831	2.41	2,002.71	104	86,424	140	116,340		1.51		
	Indian Point #2, N. Y.	2	873	2.41	2,103.93	104	90,792	124	108,252	1.79	1.51		
	Indian Point #3, N. Y.	2	965	2.75	2,653.75	142	137,030	165	159,225		1.51		
	Burlington #1, N. J.	2	999	2.75	2,747.25	142	141,858	139	138,861		1.51		
	Peach Bottom #2, Penn.	2	1065	2.75	2,928.75	142	151,230	130	138,450		1.51		
	Totals		6923 (769)		16,780.69		928,834		993,948			2.42/134	1.79/144 1.51/144

			(1)	(2)	(3)	(4)	(5)	(6)	(7)	(8)	(9)	(10)	(11)
73-77	Shoreham, N. Y.	2	540	1.75	945.00	137	73,980	185	99,900		1.51		
	Keyport, Penn.	2	800	1.75	1,400.00	137	109,600	125	100,000		1.51		
	Milliken, N. Y.	2	829	2.41	1,997.89	104	86,216	157	130,153		1.51		
	Burlington #2, N.J.	2	993	2.75	2,730.75	142	141,006	122	121,146		1.51		
	Peach Bottom #3, Penn.	2	1065	2.75	2,928.75	142	151,230	117	124,605		1.51		
	Totals		4227		10,002.39		562,032		575,804			2.37/133	1.51/136
58-62	Dresden #1, Ill.	3	200	2.33		163		255		4.45			
	Totals		200									2.33/163	4.45/255 4.45/255
63-67	Piqua, Oh.	3	11	1.77	19.47	194	2,134	727	7,997				
	Big Rock Pt., Mich.	3	70	3.15	220.50	145	10,150	386	27,020	10.23			
	Totals		81		239.97		12,284		35,017			2.96/152	10.23/432 10.23/432
68-72	La Crosse, Wisc.	3	50	2.14	107.00	292	14,600	360	18,000				
	Point Beach #1, Wisc.	3	455	1.66	755.30	157	71,435	132	60,060		1.52		

Table C.10 *Continued*

			(1)	(2)	(3)	(4)	(5)	(6)	(7)	(8)	(9)	(10)	(11)
	Point Beach #2, Wisc.	3	455	1.66	755.30	157	71,435	125	56,875		1.52		
	Kewaunee, Wisc.	3	527	1.66	874.82	157	82,739	116	61,132		1.52		
	Dresden #2, Ill.	3	715	1.68	1,201.20	154	110,110	106	75,790	2.11	1.51		
	Dresden #3, Ill.	3	715	1.68	1,201.20	154	110,110	110	78,650	2.04	1.51		
	Palisades, Mich.	3	700	1.68	1,176.00	154	107,800	146	102,200	1.40	1.51		
	Quad Cities #1, Ill.	3	715	1.68	1,201.20	154	110,110	112	80,080	2.00	1.51		
	Quad Cities #2, Ill.	3	715	1.68	1,201.20	154	110,110	112	80,080	2.00	1.51		
	Zion #1, Ill.	3	1050	1.65	1,732.50	116	121,800	156	163,800		1.51		
	Bridgeman #1, Mich.	3	1100	1.71	1,881.00	152	167,200	136	149,600		1.51		
	Totals		7197 (654)		12,086.72		1,077,449		926,267			1.68/150	1.40/129 1.51/129
73-77	Bailly, Ind.	3	515	1.66	854.90	157	80,855	177	91,155		1.52		
	Zion #2, Ill.	3	1050	1.65	1,732.50	116	121,800	146	153,300		1.51		
	Bridgeman #2, Ind.	3	1100	1.71	1,881.00	152	167,200	136	149,600		1.51		
	Totals		2665		4,468.40		369,855		394,055			1.68/139	1.51/148

			(1)	(2)	(3)	(4)	(5)	(6)	(7)	(8)	(9)	(10)	(11)
58-62													
63-67	Elk River, Minn.	4	22	3.24	71.28	259	5,598	455	10,010	11.01			
	Pathfinder, S. D.	4	59	3.16	186.44	175	10,325	356	21,004	10.76			
	Totals		81		257.72		16,023		31,014			3.18/198	10.76/383 10.76/383
68-72	Monticello, Minn.	4	472	2.17	1,324.24	114	53,808	169	79,768		1.52		
	Ft. Calhoun, Neb.	4	450	2.70	1,215.00	125	56,250	156	70,200		1.52		
	Prairie Is. #1, Minn.	4	550	2.17	1,193.50	114	62,700	178	97,900		1.52		
	Totals		1472 (491)		3,132.74		172,758		247,868			2.33/117	3.46/158[1] 1.52/158
73-77	Prairie Is. #2, Minn.	4	550	2.17	1,193.50	114	62,700	155	85,250		1.52		
	Cooper, Neb.	4	778	2.06	1,602.68	175	136,150	94	73,132		1.51		
	Totals		1328		2,796.18		198,850		158,382			2.11/150	1.51/119
58-62													
63-67													

Table C.10 *Continued*

			(1)	(2)	(3)	(4)	(5)	(6)	(7)	(8)	(9)	(10)	(11)
68-72	Robinson #2, S. C.	5	663	1.54	1,021.02	129	85,527	145	96,135	2.41	1.57		
	Turkey Pt. #3, Fla.	5	722	1.54	1,111.88	129	93,138	139	100,358		1.52		
	Turkey Pt. #4, Fla.	5	722	1.54	1,111.88	129	93,138	139	100,358		1.52		
	Surry #1, Va.	5	783	1.54	1,205.82	129	101,007	166	129,978		1.51		
	Surry #2, Va.	5	783	1.54	1,205.82	129	101,007	138	108,054		1.51		
	Crystal River #3, Fla.	5	825	1.54	1,270.50	129	106,425	133	109,725		1.51		
	Ocones #1, S. C.	5	874	2.46	2,150.04	124	108,376	98	85,652		1.51		
	Ocones #2, S. C.	5	874	2.46	2,150.04	124	108,376	98	85,652		1.51		
	Totals		6246 (781)		11,227.00		796,994		815,912			1.80/128	2.41/131 1.51/131
73-77	Surrey #3, Va.	5	800	1.54	1,232.00	129	103,200	135	108,000		1.51		
	Calvert Cliffs #1, Md.	5	848	2.46	2,086.08	124	105,152	139	117,872		1.51		
	Calvert Cliffs #2, Md.	5	848	2.46	2,086.08	124	105,152	124	105,152		1.51		
	Ocones #3, S. C.	5	874	2.46	2,150.04	124	108,376	105	91,770		1.51		
	Totals		3370		7,554.20		421,880		422,794			2.24/125	1.51/125
58-62													

			(1)	(2)	(3)	(4)	(5)	(6)	(7)	(8)	(9)	(10)	(11)
63-67													
68-72	Brown's Ferry, Ala.	6	1065	1.68		126		108					
	Brown's Ferry #1, Ala.	6	1065	1.68		126		108		0.46			
	Brown's Ferry #2, Ala.	6	1065	1.68		126		108		0.46			
	Brown's Ferry #3, Ala.	6	1065	1.68		126		108	115,020				
	Totals		4260									1.68/126	0.46/108 0.46/108
73-77													
58-62													
63-67													
68-72													
73-77	Ark. Power & Light, Ark.	7	800	1.88		73		175			1.52		
	Total		800									1.88/73	1.52/175

Table C.10 *Continued*

			(1)	(2)	(3)	(4)	(5)	(6)	(7)	(8)	(9)	(10)	(11)
58-62													
63-67													
68-72													
73-77	Fort St. Vrain, Col.	8	330	1.97		95		209					
	Totals		330									1.97/95	/209
58-62													
63-67	Humboldt Bay, Cal.	9	69	3.47	239.43	103	7,107	319	22,011	2.31			
	N. P. R., Wash.	9	790	2.91	2,298.90	102	80,580	399	315,210				
	San Onofre, Cal.	9	430	3.08	1,324.40	116	49,880	202	86,860	2.79	1.60		
	Totals		1289		3,862.73		137,567		424,081			3.00/107	2.76/329 1.60/329
68-72	Malibu, Cal.	9	462	3.08	1,422.96	116	53,592	199	91,938	1.15	1.60		
	El Diablo, Cal.	9	1060	2.91	3,084.60	102	108,120	142	150,520		1.51		
	Totals		1522		4,507.56		161,712		242,458			2.96/106	1.51/159 1.51/159

		(1)	(2)	(3)	(4)	(5)	(6)	(7)	(8)	(9)	(10)	(11)
73-77	Rancho Seco, Cal. 9	800	2.91		102		168			1.51		
	Totals	800									2.91/102	1.51/168

1 Used Nine Mile Point nuclear fuel cost. This was 68-72 plant closest in size to PR 4 average with fuel cost data available.

average with the requisite fuel figure. The fuel cost figure for the selected plant (Nine Mile Point, New York, as it turned out) was then used as the Region 4, 1968–1972 cell entry. The complete series for Variant A is to be found in column 8 of Table C.1.

The basic source of our second nuclear fuel price series (Variant B) was a 1966 General Electric price list. The general method here was to assign to every nuclear plant the fuel price for a General Electric plant of comparable size. This series, also incomplete, is given in column 6 of the fuel cost worksheets. To generate the cell series given in column 9, Table C.1, the following procedure was used: in those cells where there was at least one General Electric price entry, we used the procedure for the Variant B figures; where, however, there were no General Electric price entries in a cell, we used the corresponding Variant A cell figure.

The choice between these two was deferred to a final inspection. The result was that Variant B was selected, since — as a composite, in effect of two series — it showed the most stability from cell to cell and time period to time period.

The Total Capacity Equation

The best least-squares regression equation obtained to explain total capacity demanded was

$$\log Q = -7.046 - 1.240 \log P_e + 0.855 \log \frac{Y}{N} + 0.914 \log N,$$

where

Q = total steam-electric capacity (fossil and nuclear) demanded,
P_e = price of electricity index,
Y/N = per capita constant dollar personal income,
N = population.

The particular task here will be, as before, to explain in some detail the specific sources of the dependent and independent variables series used. Table C.11 is a summary table of these results.

Total Capacity

This is the same variable as Q in the nuclear share equation, and hence need not be discussed here.

Electricity Price Index

The first problem to be confronted here was that of obtaining a reasonable annual and regional electricity price series. The source finally selected was the Edison Electrical Institute's *Statistical Yearbook of the Electric Utility Industry*. The specific price measure decided upon

Table C.11 Total Capacity Equation Observations

Period (1)	PR (2)	Total Addition to S. E. Capacity MW_e (3)	Electricity Price Index (4)	Population (5)	Per Capita Pers. Income (1957 - 1959 dollars) (6)
1958-1962	1	1,388	1.47911	10,527	2,408
	2	7,101	1.17772	34,287	2,555
	3	10,769	1.01453	36,286	2,367
	4	2,805	1.29096	15,419	2,053
	5	7,676	1.04299	26,095	1,819
	6	4,543	0.49245	12,083	1,464
	7	6,247	1.03892	17,023	1,793
	8	1,472	0.89316	6,913	2,080
	9	3,675	0.78747	20,490	2,600
1963-1967	1	1,732	1.34645	11,151	2,921
	2	5,039	1.08115	36,487	3,024
	3	7,744	0.94936	38,254	2,910
	4	3,608	1.16673	15,858	2,560
	5	11,410	0.91969	28,762	2,305
	6	5,271	0.50149	12,825	1,872
	7	9,294	0.91912	18,500	2,179
	8	2,072	0.86321	7,686	2,442
	9	7,918	0.72628	23,815	3,029
1968-1972	1	7,671	1.19322	11,699	3,543
	2	19,332	0.97816	38,325	3,579
	3	24,540	0.78629	39,933	3,578
	4	10,906	1.03196	16,203	3,192
	5	23,879	0.83325	31,210	2,921
	6	20,065	0.35712	13,419	2,394
	7	17,124	0.78475	19,860	2,648
	8	8,028	0.70260	8,455	2,867
	9	8,348	0.65924	26,178	3,529

was the ratio of total revenue to total sales (in kilowatt hours) from sales to all consumers (industrial, commercial, residential, and other), both series being provided by the *Yearbook* on a regional basis. Thus there was an electricity price for every region and every year from 1945 to 1965 (see Table C.12).

Table C.12 Electricity Prices, 1945–1956, by Regions

	Price of Electricity in Region 1								
2.466	2.520	2.521	2.605	2.679	2.554	2.522	2.561	2.524	2.608
2.545	2.554	2.608	2.648	2.557	2.547	2.517	2.498	2.479	2.420
2.360									
	Price of Electricity in Region 2								
1.867	1.942	1.923	1.963	2.035	1.990	1.932	1.982	1.991	2.057
1.996	2.011	2.049	2.111	2.072	2.028	2.027	2.001	1.961	1.920
1.895									
	Price of Electricity in Region 3								
1.792	1.900	1.838	1.861	1.968	1.912	1.867	1.891	1.875	1.957
1.753	1.678	1.708	1.745	1.728	1.747	1.743	1.729	1.699	1.680
1.664									
	Price of Electricity in Region 4								
2.265	2.334	2.291	2.333	2.410	2.373	2.353	2.377	2.384	2.425
2.391	2.362	2.349	2.364	2.326	2.223	2.233	2.193	2.156	2.101
2.045									

Price of Electricity in Region 5

1.698	1.747	1.714	1.757	1.880	1.828	1.809	1.816	1.811	1.853
1.827	1.820	1.849	1.849	1.803	1.796	1.761	1.732	1.703	1.659
1.612									

Price of Electricity in Region 6

1.026	1.083	1.088	1.117	1.171	1.155	1.154	1.100	1.009	.922
.811	.791	.796	.823	.829	.848	.869	.888	.876	.882
.879									

Price of Electricity in Region 7

1.854	2.021	1.993	2.005	2.041	1.966	1.883	1.862	1.852	1.851
1.837	1.808	1.801	1.821	1.791	1.789	1.810	1.785	1.730	1.678
1.611									

Price of Electricity in Region 8

1.664	1.698	1.653	1.650	1.639	1.662	1.655	1.650	1.604	1.585
1.542	1.496	1.527	1.562	1.577	1.538	1.520	1.524	1.541	1.526
1.513									

Price of Electricity in Region 9

1.246	1.337	1.281	1.251	1.253	1.246	1.232	1.281	1.315	1.305
1.269	1.265	1.285	1.380	1.355	1.356	1.369	1.364	1.342	1.306
1.273									

Next we transformed this series into a price index. The formula used was

$$I = \frac{P_e/1.710}{\mathrm{WPI}/100}, \qquad (8)$$

where

P_e = price of electricity (regional),
1.710 = national electricity price in 1958,
WPI = wholesale price index (base period = 1957–1959).

Two problems then remained. First a five-year index had to be built from the annual index series, so as to fit price into the general regression schema. The middle-year figure was used for this index. Thus the 1960 regional price indices were used as the cell entries for the 1958–1962 period, and the 1965 and 1970 indices as the entries for the 1963–1967 and 1968–1972 periods, respectively. Second, the 1970 index had to be constructed. The Edison Electric Institute data did not extend beyond 1966, hence we had to improvise. What in fact was done was to calculate regional log linear least-squares trend lines based on the 21 index observations from 1945 to 1965 and then project those lines to 1970. These same equations, as will be mentioned below, were used later to make much longer-range price forecasts. (The equations are presented in the fuel cycle worksheets.)

Hence the first 18 entries in column 4, Table C.11 are simply transformations of 1960 and 1965 price data given by the Edison Electric Institute's *Yearbook* while the last nine are forecasts of the index for the year 1970.

Population

Here the Bureau of the Census data and forecasts were used. Table C.13, columns 2, 3, and 4 give the relevant population observations with notes explaining the specific sources. This table is the immediate source for column 5, Table C.11.

One will note, by the way, if he consults P25, No. 388 that there are several alternative estimates there for the 1970 regional populations. Our choice was necessarily an arbitrary one since the Bureau itself claims that none of the alternatives can be favored *a priori* over any of the others.

Personal Incomes

Table C.14 should be self-explanatory. The relevant columns are, of course, the last three.

Forecasting Total Capacity Beyond 1970

The total capacity equation previously discussed was used as the forecasting instrument. The problem to be discussed here is the fore-

Table C.13 Population, Data and Projections

	PR (1)	1960[1] (2)	1965[1] (3)	Series 1b 1970[3] (4)	1975[3] (5)	1980[3] (6)	1985[3] (7)
1	New England	10,527	11,151	11,699	12,471	13,416	14,469
2	Middle Atalntic	34,287	36,487	38,325	40,747	43,549	46,525
3	E. N. Central	36,286	38,254	39,933	42,534	45,861	49,713
4	W.N. Central	15,419	15,858	16,203	16,896	17,854	18,991
5	South Atlantic	26,095	28,762	31,210	34,232	37,705	41,423
6	E. S. Central	12,083	12,825	13,419	14,228	15,154	16,109
7	W. S. Central	17,023	18,500	19,860	21,484	23,289	25,130
8	Mountain	6,913	7,686	8,455	9,398	10,513	11,730
9	Pacific	20,490[2]	23,815[2]	26,178[2]	29,762[2]	33,743[2]	38,229[2]

Notes

1 Current Population Reports: Population Estimates, "Estimates of Population by States: July 1, 1966", Series P-25, No. 380, November 24, 1967, Table 3, Bureau of the Census.

2 Excludes Alaska and Hawaii.

3 C. P. R.: Population Estimates, "Summary of Demographic Projections", Series P-25, No. 388, March 14, 1968, Table 16, Bureau of the Census.

casting of the *independent* variables to be substituted into the equation. A summary of these results, including the forecast capacities, is given in Tables C.16 to C.19.

Population

The Bureau of the Census provides regional population forecasts for the years 1970, 1975, 1980, and 1985 and national forecasts for 1990, 1995, 2000, and 2005. The figures desired were annual regional figures for the years 1987, 1992, 1997, and 2002, these being the mid-years of four consecutive five-year periods beginning in 1985.

Table C.14 Total and Per Capita Personal Income, by Regions

	Total Per. Income $1,000's		Per Capita Pers. Income		Constant Dollar Per Capita Personal Income		
PR (1)	1960[1] (2)	1965[1] (3)	1960[5] (4)	1965[5] (5)	1960[3] (6)	1965[4] (7)	1970(proj.)[6] (8)
1	25,532	33,383	2,425	2,994	2,408	2,921	3,543
2	88,204	113,116	2,573	3,100	2,555	3,024	3,579
3	86,490	114,109	2,384	2,983	2,367	2,910	3,578
4	31,871	41,609	2,067	2,624	2,053	2,560	3,192
5	47,809	67,951	1,832	2,363	1,819	2,305	2,921
6	17,821	24,610	1,475	1,919	1,464	1,872	2,394
7	30,743	41,304	1,806	2,233	1,793	2,179	2,648
8	14,482	19,235	2,095	2,503	2,080	2,442	2,867
9	53,646[2]	73,949[2]	2,618[2]	3,105[2]	2,700[2]	3,029[2]	3,529[2]

Notes

1 Survey of Current Business, volume 46, number 8, August 1966, page 12.

2 Excludes Alaska and Hawaii.

3 Column 2 ÷ 1960 WPI (= 100.7, 57-59 = 100).

4 Column 3 ÷ 1965 WPI (= 102.5, 57-59 = 100).

5 Columns 2 and 3 ÷ corresponding population figure in Table 13.

6 Assume average regional rate of growth of CDPCPI from 1960 to 1965 continues from 1965 to 1970.

The first and most important step was to make regional estimates for the years 1990, 1995, 2000, and 2005. To do this we calculated the share of each region in the total national population in 1980 and 1985 and then subtracted the former from the latter. (See columns 2, 3, and 4, Table C.15.) The assumption was then made that this "trend" would continue through the year 2005. Thus, if Region A's share was X per cent in 1980 and Y per cent in 1985 we assumed its 1990 share was $[Y + (Y - X)]$ per cent, its 1995 share was $[Y + 2(Y - X)]$ per cent, etc. The results of these calculations are set out in columns 5 to 8 of Table C.15. Finally, these regional percentages were applied to the national forecasts provided by the Bureau of the Census to give the five-year regional forecasts found in columns 9 to 12.

Table C.15 Forecasting Regional Populations 1990, 1995, 2000, and 2005

PR (1)	1980 Share % (2)	1985 Share % (3)	Δ (4)	1990 Share % (5)	1995 Share % (6)	2000 Share % (7)	2005 Share % (8)	Population in 1990 (9)	Population in 1995 (10)	Population in 2000 (11)	Population in 2005 (12)
1	5.56	5.52	-0.04	5.48	5.44	5.40	5.36	15,700	16,855	18,143	19,590
2	18.06	17.74	-0.32	17.42	17.10	16.78	16.46	49,908	52,981	56,377	60,080
3.	19.02	18.95	-0.07	18.88	18.81	18.74	18.67	54,091	58,279	62,962	68,100
4	7.41	7.24	-0.17	7.07	6.90	6.73	6.56	20,256	21,378	22,611	23,980
5	15.64	15.79	+0.15	15.94	16.09	16.24	16.39	45,668	49,852	54,563	59,850
6	6.29	6.14	-0.15	5.99	5.84	5.69	5.54	17,161	18,094	19,117	20,220
7	9.66	9.58	-0.08	9.50	9.42	9.34	9.26	27,218	29,186	31,380	33,810
8	4.36	4.47	+0.11	4.58	4.69	4.80	4.91	13,122	14,531	16,127	17,910
9[2]	14.00	14.57	+0.57	15.14	15.71	16.28	16.85	43,376	48,674	54,697	61,550
								286,501[1]	309,830[1]	335,977[1]	365,254[1]

Notes

1 1990 - 2005 National forecasts from CPR "Population Estimates" series P-25 #359, February 20, 1967.

2 Excludes Alaska and Hawaii.

The second and last step was to interpolate logarithmically between these predictions to get forecasts for the particular years desired. The results here are to be found in Table C.16.

Table C.16 Regional Population Forecasts — 1987–2002

	Population (in millions)			
	1987	1992	1997	2002
Region 1	14.9	16.1	17.3	18.7
Region 2	47.8	51.1	54.3	57.8
Region 3	51.4	55.7	60.1	64.9
Region 4	19.4	20.7	21.9	23.1
Region 5	43.1	47.3	51.7	56.6
Region 6	16.5	17.5	18.5	19.6
Region 7	25.9	27.9	30.0	32.3
Region 8	12.3	13.7	15.1	16.8
Region 9	40.2	45.4	50.9	57.3

Personal Income

Here again a steady state growth assumption was employed. The average *national* rate of growth of per capita constant dollar personal income between 1955 and 1965 was 3.0 per cent. This calculation with all due acknowledgments can be found in Appendix C.2. We then assumed that income in all regions would grow at this rate constantly throughout the time period in question. The results of this calculation can be found in Table C.17.

Table C.17 Regional Per Capita Personal Incomes — 1987–2007 (1957–1959 dollars)

	1987	1992	1997	2002
Region 1	$5760	$6645	$7667	$8844
Region 2	5819	6712	7744	8934
Region 3	5817	6710	7742	8932
Region 4	5188	5987	6407	7968
Region 5	4748	5479	6320	7292
Region 6	3892	4490	5180	5977
Region 7	4305	4967	5729	6610
Region 8	4661	5377	6203	7157
Region 9	5737	6619	7636	8809

Electricity Prices

The electricity price forecasting equations, whose origins were discussed above, were used here to generate the requisite forecasts; since the equations are given in the fuel cycle worksheets, this is pure arithmetic and will not be discussed any further here. The forecasts themselves are shown in Table C.18.

Table C.18 Regional Index of Electricity Prices — 1987–2002

	(index based on 100 in 1960)			
	1987	1992	1997	2002
Region 1	84.4	76.2	68.9	62.3
Region 2	72.7	66.7	61.2	56.1
Region 3	51.4	45.3	40.0	35.4
Region 4	69.3	61.7	54.9	48.9
Region 5	58.8	53.1	47.9	43.2
Region 6	19.7	16.5	13.8	11.7
Region 7	49.6	43.4	37.9	33.2
Region 8	46.4	41.1	36.4	32.2
Region 9	50.7	46.9	43.5	40.3

In summary, the effects of all of these assumptions and forecasts are found in the forecasts of total capacity given in the text of Chapter 3. In future research, the accuracy of these first forecasts will be obvious, and the data and equation forms can be revised to increase this accuracy.

Appendix D

Characteristics of Breeder Reactor Production Functions

1. The Liquid Metal Fast Breeder Reactor

All reactors, to this point in time, have been built around a core made up of fuel rods in which uranium or plutonium fission takes place. The core may be cylindrical or pancaked in shape, and the number and size of rods can vary with core temperature and the means for removing the heat energy. In "fast breeder reactors," as contrasted with "thermal reactors," the lack of moderating material in the core results in reduced probability of fissioning of uranium (U^{233}, U^{235}), but increased probability of nonfission capture of neutrons by U^{238} so as to "breed" fissionable plutonium Pu^{239}. The bred plutonium, or an original inventory of the same material, then provides heat energy.

Liquid sodium is one coolant that can be used to transfer the heat from fast fission from the core to a steam source for power generation. Sodium is molten at 200°F and boils at 1,600°F, so that a wide range of heat-transfer temperatures is possible at environmental pressures. It has excellent heat absorption characteristics, but is corrosive to many metals and reacts so strongly to water that the combination is explosive under pressure.

As a result of these characteristics, the flow diagram for a sodium-cooled fast reactor is complex. The heat transfer subsystem from the core to the steam generator is divided into a number of independent and self-contained loops so that the failure of sodium flow at one location will not disrupt the entire transportation network. Most loops have back-up and by-pass routes, as well, so that any malfunction resulting from sodium corrosion is isolated and separated from the continued operation of other components in that loop. Typically, the heat is routed by sodium from the core through primary transfer loops to three or more secondary sodium loops, and then transferred from there to a number of separate steam generators; each primary or secondary sodium loop is self-contained and the heat exchanger is made up of a shell of one material in which is contained the tube of the second material.

There have been four major design studies of 1980 LMFBR's with capacity close to 1,000 MW_e, and a number of article-length analyses either of smaller reactors or of particular design variations.[1] Each of the major designs has a set of elaborate heat transfer loops — in particular, six separate sodium loops for the transfer of heat to two separate steam generators. But no particular design has a specific requirement for a core configuration or even a fuel rod; the configuration and arrangement of rods, and their life cycle from fabrication to removal from the core, do not differ in kind from those in the major designs of other types of reactors.[2] The dimensions of these elaborate capital systems K and less elaborate fuel requirements F can be found by cross reference in the four designs, and by comparison with earlier designs of smaller liquid metal breeders.

[1] The four design studies for a large LMFBR sought to provide 1,000 MW_e of capacity in each case. Some of the designs involved higher levels of thermal efficiency than others, so that the thermal megawatts of capacity varied around 2,500 MW_t. The four studies are:

Large Fast Reactor Design Study, Allis-Chalmers Rept. ACNP-64503 (January, 1964).

Liquid Metal Fast Breeder Reactor Design Study, Combustion Engineering Rept. CEND-200 (January, 1964).

M. C. McNally *et al.*, *Liquid Metal Fast Breeder Reactor Design Study*, General Electric Rept. GEAP-4418 (January, 1964).

Liquid Metal Fast Breeder Reactor Design Study, Westinghouse Electric Rept. WCAP-3251-1 (January, 1964).

These are referred to as the "major designs"; but there are others of importance of interest as to particular parameters. The study by K. P. Cohen and G. L. O'Neil, "Safety and Economic Characteristics of a 1,000 MW_e Fast Sodium-Cooled Reactor Design" (ANL 6700, 1965) contributes to the evaluation of system reliability. The study by W. Hafele, D. Smidt, and K. Wirtz, "The Karlsruhe Reference Design of a 1,000 MW_e Sodium-Cooled Fast Breeder Reactor," (*ibid.*) also contributes to the analysis of accidents which disrupt system reliability. Studies of smaller liquid metal reactors by General Electric contribute to the analysis of scale: cf. H. E. Dodge, *et al.*, *Conceptual Design of a 565* MW_e *Fast Ceramic Reactor*, General Electric Rept. GEAP-4226 (April, 1963), and K. N. Horst, *et al.*, *Core Design Study for a 500* MW *Fast Oxide Reactor*, General Electric Rept. GEAP-3721 (December 28, 1961).

[2] This is not to deny the existence of different configurations in the design of the fuel core. For example, there are striking differences among the four studies themselves, as shown in the view of the Reactor Engineering Division, Chicago Operations Office, U.S. Atomic Energy Commission, *An Evaluation of Four Design Studies of a 1,000* MW_e *Ceramic-Fueled Fast Breeder Reactor*, Rept. COO-279 (December 1, 1964), in the diagram of reactor core arrangements of the four design studies on the first page. But these core configurations are not specific to the liquid metal fast reactor; in fact, the configurations in the gas- and steam-cooled reactors discussed below are more similar to those of Combustion Engineering Corporation and General Electric than these two designs are to the Westinghouse and Allis-Chalmers designs for the liquid metal reactor.

The General Electric study, and that of Allis-Chalmers, indicate capital expenditures in great detail.[3] Consider capital K to be items of equipment for the production of heat energy in the form of steam at 1,000°F and 2,500 psia, where these items include all systems in the reactor and for the transfer of heat to sodium intermediate loops and then to the steam generator. Then the relevant equipment includes (a) the reactor vessel and internals, (b) primary sodium pumps, drives, and piping, (c) intermediate heat exchangers, (d) secondary sodium pumps, drives, and piping, and (e) final heat exchangers including the steam generating system. But it does not include the turbine generator building, the generator unit, and its accessory electrical equipment. The total of expenditures for the relevant equipment, termed $\sum PQ$ for component prices P multiplied by respective quantities Q, comes to $\$84.6 \cdot 10^6$ and the total number of components $\sum Q$ is 467.[4] Not all components are of the same size, nor do they involve the same amount of fabrication and engineering of the same metals; weighting each by its price — so that those with higher prices are assumed to be larger — the average size of a component is $\sum PQ/\sum P = 84.6(10^6)/46.9(10^6) =$ 1.80 units of capital. Then total capital comes to $(\sum PQ/\sum P)\sum Q =$ $(1.80)(467) = 841$ units.

The quantity of fuel required for 1,000 MW_e is investigated in all four design studies. Total fuel mass is divided between "the core" and "the blanket" surrounding the center of fission. But the latter has qualities and functional use different from the former, with the plutonium in the core used both to produce heat energy and neutrons for converting the blanket uranium to more plutonium. The plutonium creates and sustains fission and is necessary in a fast environment for breeding. Capacity to produce thermal megawatts is determined, then, by the

[3] The implications of particular constraints in these studies cannot be known with any exactitude. But it would seem quite likely that they stand in the way of finding the least-cost combination K_1, F_1 for given MW_1 even with the factor prices stated in the Atomic Energy Commission *Guide to Nuclear Power Cost Evaluation*, Vol. 5, *Production Costs*, Kaiser Engineers Rept. TID 7025 (March 15, 1962). Both the constraints on capital and on fuel set the levels of utilization of these two inputs, rather than allowing the ratio of marginal products of the inputs to be equated to the ratio of factor prices. Of more concern is the possibility that the observation of F_1 is not the minimum amount required to produce MW_1 with any stated K_1 — the specification of maximum core outlet temperature may require more F_1 than is necessary. Then variation of the K, F combination would not lead to an approximation of the least-cost combination of inputs, because the observed combination may include too much of both inputs. This possibility is left open here, although the analysis proceeds as if it did not hold.

[4] Cf. General Electric Rept. GEAP-4418, Section 2.8, "Economics Data." The number of components is estimated from counting items of equipment in Table 2.8.2.1, "Reactor Equipment Cost Summary," and — where the table is incomplete — in the plant layout charts and diagrams shown throughout the report.

kilograms of fissile plutonium in place at all times during the lifetime of the capital equipment. The initial plutonium loading M_0 provides this capacity when the equipment is installed (since this amount is capable of producing 1,000 MW_e). But this loading lasts only two to three years, and capital is in operable condition twenty to thirty years so that the inventory required to provide lifetime capacity is the present or initial sum of all required future loadings — that is, the sum of the present discounted values of the volumes required to be installed for continuous capacity over the equipment lifetime.

The time perspective for planning productive capacity is from the initial installation of equipment over the lifetime of that equipment. The planner has to consider capital and fuel in the same dimension — that quantity of each input factor required to provide 30-year capacity to produce a certain amount of thermal energy. The capital capacity is indicated by the equipment installed during construction along with certain items which have to be replaced before the 30-year lifetime is complete. The fuel required is the sum of annual loadings of fuel and blanket uranium, designated $\sum N_i$ for $i = 1$ to $i = 30$. But these annual loadings do not all take place at the time of the initial installation of the equipment; rather they can be postponed with consequent savings of investment capital shown by the annual rate of return r on such invested capital. Each loading N_i can be discounted by $(1 + r)^i$, so that fuel inventory at the time plans are made for construction equals

$$\textstyle\sum_i N_i\,(1 + r)^{-i}.$$

This amount does not yet take account of the gains from breeding. During each inventory life, B units of fissionable material are produced as a ratio of those consumed; then $B^{1/t}$ units are produced annually given that t = length of life of an inventory of core fuel and $(B^{1/t} - 1)$ of that inventory can be removed as excess each year. At the end of the first year, the necessary core inventory is $N_0 - N_0(B^{1/t} - 1)$ and the present value of this amount is $N_0 - N_0(B^{1/t} - 1)/(1 + r)$. At the end of the second year, the fuel core has again increased by $(B^{1/t} - 1)$ and a three-year cycle requires $F = N_0 - N_0(B^{1/t} - 1)/(1 + r) - N_0(B^{1/t} - 1)/(1 + r)^2$. The fuel requirements for the 30-year reactor life are $F = N_0[1 - (B^{1/t} - 1) \sum (1 + r)^{-i}]$.

According to the General Electric design analysis, 2,357 kg of fissile plutonium are installed on the day the plant ceases to be a construction project (or, according to the Allis-Chalmers design, 2,910 kg are installed).[5] This volume of fuel is to be loaded again and again over the reactor lifetime, so that it is always in place; but at the same time the

[5] The differences in fuel inventories can be accounted for in terms of temperature and pressure conditions in the core, which utilize fuel more intensively in the first case, and less intensively in the second case. If the technologies assumed in the

breeding of new plutonium provides more fuel in place than this amount installed.[6] For the General Electric case, the value of fissionable fuel in the core increases at the rate of $(B^{1/t} - 1) = 0.081$ per year. The present value of all future loadings declines at a rate of at least 10 per cent in each year from the time at which the initial loading takes place, given that the minimum opportunity costs for fuel in use are at least this great, so that loadings in year t are equivalent to the initial loading discounted by 10 per cent for each intervening year. The sum of 30 loadings over the reactor lifetime is approximately $F = (2{,}357)[(1 - 0.081) \sum_1^{30} (1.1)^{-t}]$ or 558 kg. Then $K = 841$ and $F = 558$ is an approximate single observation of capacity and capacity input requirements for the general LMFBR production function.

A trade-off of fuel for capital is involved in changing the "specific power" of the system, as measured in energy kilowatts per kilogram of fissile plutonium.[7] Higher specific power can be attained by increasing the thermal conductivity of the fuel, or the fuel and coolant temperatures; either course of action adds to thermal stress and corrosion in the core assembly and to the capital items in the shielding and heat transfer systems. Both reduce the fuel inventory. The requirements for more capital at higher power are shown in the Combustion Engineering study in the increase in fuel assembly components and the decrease in the radius and length of each component — where, as a first approximation, the increased fabrication implied by more and smaller rods is equivalent to an increase in components. The corrosion and deterioration due to increased temperatures and pressures require higher quality components or more frequent replacement of components in reactor vessels and internals. In the Combustion Engineering study, these new requirements altogether are equivalent to a 11.3 per cent increase in these components given a 100 per cent increase in specific power.[8] The

design studies are the same, then the two values of F are observations for fuel and capital in different combinations producing the same amount of output. That is, the design studies show F_1, F_2 for $MW_t = \lambda = f(K, F)$, an isoquant of minimum amounts of capital and fuel to produce capacity set at λ.

[6] Cf. *A Study of a Gas-Cooled Fast Breeder Reactor: Initial Study, Core Design Analysis and System Development Program, Final Summary Report*, General Atomics Division of the General Dynamics Corporation, Rept. GA-5537 (August 15, 1964), Section 12.1, "Fuel Cycle Cost Equation," pp. 163–167.

[7] The "specific power" of a reactor design is defined as thermal megawatts of capacity divided by the equilibrium fissile fuel inventory. In economic terms specific power is the average productivity of fuel MW/F.

[8] Figure IV-27 in the Combustion Engineering study shows that a 100 per cent increase in specific power over the base value of 175 kW/kg causes an increase in capital costs of 0.19 mill/kWh. This is 11.3 per cent of the imputed costs of 1.68 mills with an 80 per cent load factor and all factor prices unchanged. Then an increase in costs must take place from increased quantities of inputs equal to 11.3 per cent of the original value of these inputs.

reduction in fuel inventory from increased specific power is shown in both the General Electric and Allis-Chalmers studies. By "doubling the core power density and specific power, simultaneously reducing the number of batches to keep the same refueling schedule — [and accepting] a penalty in the physics area . . ." the input plutonium content is used up more completely and thus the inventory to produce a given capacity is reduced; the net effect in the General Electric design is that fuel inventory is reduced approximately 9.5 per cent.[9]

The trade-off of fuel for capital $\partial K/\partial F$ can be shown by the results of these changes to reach higher specific power. If the increase in fuel assemblies and reactor internals as shown by Combustion Engineering is applied to the number of units of capital shown in the General Electric design, then 28.4 more units of capital are required to increase power by 100 per cent. This reduces fuel by 51.8 units, so that $\partial K/\partial F = 0.55$ or 55 units of capital are added for each 100 unit reduction in fuel.

The capital and fuel requirements at different thermal energy outputs have not been investigated as part of any one design study. General statements have been made on the advantages of large scale: The Atomic Energy Commission design handbook is based on output increasing at a faster rate than inputs, so that larger plants have lower capital/output and fuel/output ratios than smaller plants; but the Westinghouse design study, in considering both small and large versions of a liquid metal reactor, shows no decrease in inputs per megawatt of capacity at large scale.[10] The only detailed studies of different sizes of the same reactor are two General Electric designs, those for the 1,000 MW_e liquid metal fast reactor and a 500 MW_e reactor based on the same coolant systems and core configuration. The conceptual design for the first indicates, as shown above, $K = 841$ and $F = 558$; the second design provides most if not all the information necessary for constructing comparable estimates of K and F for MW = 500.

The 500 MW design is based on the same technology — the study was completed in 1963, rather than 1964, but it has the same flow diagram as the larger reactor.[11] Four primary sodium pumps move the liquid coolant through the bottom and then out of the top of the reactor core to four intermediate heat exchangers which also contain sodium; the

[9] Cf. General Electric Rept. GEAP-4418, pp. 5–7. The change in inventory = $-0.074/0.569 = -9.5$ per cent (3–134).

[10] Kaiser Engineers Rept. TID 7025 (March 15, 1962), Vol. 5, *Production Costs*, Section 306. For the Westinghouse design study, cf. Westinghouse Electric Rept. WCAP-3251-1.

[11] Compare Figure 2.4 in General Electric Rept. GEAP-4418, the 1,000 MW design study, with Figure 9 in General Electric Rept. GEAP-4226, the basic 500 MW design study. The flow diagrams are the same, at least in these simplified versions of the two systems.

secondary loops then carry the heat energy to a once-through steam generator or to steam reheater loops. Such a system differs only in the number and size of loops or other components, when compared to the 1,000 MW General Electric liquid metal design, but not the techniques of heat transfer, the use of an intermediate transfer network, or the design of the reactor vessel and internals. Sodium is transported out of the reactor at 1,050°F and 26 psi at the rate of 60×10^6 lb/h in the smaller reactor, and at the same temperature and pressure — but in larger quantity, at 86×10^6 lb/h — in the larger reactor.[12] The "product," evidenced by heat energy transported by the sodium to pass-through steam generators, would seem to be the same.

The costs of units of capital, defined as $(\sum P^*Q^*)$ for the small reactor for comparability with capital $(\sum PQ)$ in the larger system described above, are $\$62.9 \cdot 10^6$ for components and construction of the system. The number of components $\sum Q^*$, and the sum of prices $\sum P^*$, are not shown in the study, but a sample of prices for reactor equipment is identical to the sample in the design for the larger reactor,[13] so that $\sum P^*$ is assumed to be equal to $\sum P$ for the larger design. Given the similarity in the designs of the two systems, the number of components should not differ either, so that both $\sum P^* \sim \sum P$ and $\sum Q^* \sim \sum Q$. The difference between systems should be in the size of components with the smaller average size occurring in the smaller reactor. Then capital $(\sum P^*Q^*/\sum P^*) \sum Q^*$ equals $(\sum P^*Q^*/\sum P)\sum Q$, which is [(62.9)/(46.9)](467) or 625 units.

The smaller reactor uses more fuel for producing a given amount of heat energy, it seems. The initial core loading of fissionable plutonium, equal to 1,200 kg, is larger per megawatt of capacity than that in the 2,500 MW_t design reactor. This smaller machine does not "breed" at as high a rate because capture of neutrons by fertile uranium is less complete in a smaller core; thus the additional fuel loadings are larger relative to the first core inventory. The burden of a low rate of breeding, in fact, shows clearly in $(B^{1/t} - 1) = 0.042$ rather than the 0.081 expected for the full-scale 2,500 MW_t system.[14] The impact of these two

[12] The basis for differences in rates of flow is in the size and number of components to achieve greater heat transfer so as to make the capacity level for the larger reactor.

[13] This is the case at least for those basic items such as steel plate and croloy for which price comparisons are possible. Cf. General Electric Rept. GEAP-4418, Table 2.8.2.1 and General Electric Rept. GEAP-4226, Table IV.

[14] As calculated from information in the General Electric analysis for the formula $B = [h(r - 1) + 1]$ with $h = 24bE_m m(1 + \alpha)/103E_f xe$, and r the breeding rate. The variables in h are b = total fissions/core fissions, E_m = maximum core burn-up, m = maximum/average power, α = capture/fission, E_f = fission energy yield, x = fast fission factor, e = fission weight/total fuel weight.

factors is large: total lifetime inventory, viewed from the day in which plant construction is begun, is 1,200[1 – 0.042(9.43)] or 729 kg.

The values of K and F for MW in the two reactors, and of $\partial K/\partial F$ for the larger reactor, provide information for a first approximation to the LMFBR production function. The two designs indicate for the equation $\beta'(\log K_1/K_2) + \psi'(\log F_1/F_2) = \log(\mathrm{MW}_1/\mathrm{MW}_2)$, that $\beta'(1{,}290) + \psi'(-1{,}161) = 2{,}534$. The trade-off of capital for fuel on specific power $\partial K/\partial F = -\psi' K/\beta' F = -55/100$ or, for the design value of 2,500 $\mathrm{MW_t}$, $\psi'(841)/\beta'(558) = 55/100$. Solving these two equations yields $\beta' = 2.9$ and $\psi' - 1.1$ or a description of the function as $\mathrm{MW} = \alpha K^{2.9} F^{1.1}$.

Even the first approximation provides a characterization of the liquid metal breeder. The extensive heat absorption ability of liquid sodium comes to fruition in the large-scale reactor to a much greater extent than in the small scale; this is shown by the very large sum of exponents, so large that a doubling of inputs increases output capacity 16 times over. Small additions of capital make large additions to thermal energy capacity, as shown by the values of the marginal product of capital in the ranges of F, K in the two design studies; this marginal product, or $\partial \mathrm{MW}/\partial K = \beta \mathrm{MW}/K - 2.9\ \mathrm{MW}/K$ increases over the range from 500 MW to 1,000 MW. If capital is added to a given level of fuel output increases by a greater amount up to technical limits set by permissible metal temperatures and pressures.

A second approximation to the liquid metal production function has to account for deviations of output from that for which the reactor is designed. The realized outputs from spaced runs with a given amount of equipment and fuel are very seldom the same, because driving the core at limit thermal conditions causes distortion in the shape of the fuel rods consequent to the energy output. There is always some chance of melting the fuel when there are random increases in temperature or transient changes in power, and the latter involve the risk of excess reactivity rendering fission uncontrollable.[15]

[15] The changes in reactivity with respect to temperature and power determine the stability of the system. Temperature-induced reactivity changes $\partial\rho/\partial T$, with ρ defined as the per cent deviation from steady-state levels of fission activity, are random events; if a temperature increase is not to increase fission, so as to increase temperature and fission once again and render the system unstable, then values of $\partial\rho/\partial T < 0$ are required. In some conceivable fast sodium-cooled reactors, increases in temperature reduce fuel and sodium density, and reduced density implies $\partial\rho/\partial T < 0$. But in other conceivable designs for this reactor with a compact core of minimum possible size fissioning at high temperature, the fast neutron energy spectrum may cause $\partial\rho/\partial T$ to have the opposite sign. That is, it is possible that (1) $\partial\rho_1/\partial T < 0$, where ρ_1 results from reduced sodium density, but the magnitude of this negative effect decreases as the compactness of any core increases; (2) $\partial\rho_2/\partial T > 0$ where ρ_2 is the effect of sodium in "degrading the neutron energy

The three studies of the large liquid metal reactor have been concerned with forced outage and other deviations from target levels of operation; necessarily decisions have been made to set lower targets so as to reduce these deviations.[16] The departures resulting from transient temperature changes which have been found acceptable by the originators of the designs are a function of the design temperature and level of operation. In $Q_n = \alpha K^{\beta'} F^{\psi}$ the constant term α takes the form $\alpha_0 e^{-x}$, where x equals the temperature variant $\Delta T/T$ times $(1 + \xi)$, with ξ the temperature-induced elasticity of reactivity $T\, \partial\rho/\rho\, \partial T$. When actual temperature T equals design temperature T^*,

$$\alpha = 2.5 \times 10^{-9} \qquad \text{and} \qquad Q_n = 2.5(10^{-9}) K^{2.9} F^{1.1},$$

the design values of inputs and capacity output. But when a transient results in a 50° increase in temperature at $T^* = 1{,}000$°F and the coefficient of reactivity is -0.005, then capacity is reduced by 5 per cent. A 25 per cent deviation in temperature substantially reduces the effectiveness of the reactor — but by much less than from a melting accident which would take place in a less conservative design. Temperature deviations in this reactor are expected to result in penalties on available capacity of small magnitude; at least the safety rod and fast shutdown systems in the General Electric design can be expected to go into effect for changes in temperatures of any importance. Then this approximation of the production function is

$$Q_n = \alpha K^{\beta'} F^{\psi} = 2.5(10^{-9}) K^{2.9} F^{1.1} e^{-(\Delta T/T)(1-0.005)}$$

for instantaneous capacity Q_n, capital K, fuel inventory F, coolant outlet temperature T, and ξ equal to -0.005 in this case.

---

spectrum" so as to increase fission from the available neutrons; (3) the size of $\partial\rho_2/\partial T$ becomes larger with larger reactor size. Thus for larger reactors with compact cores, the sum total effect may be either $\partial(\rho_1 + \rho_2)/\partial T < 0$ or $\partial(\rho_1 + \rho_2)/\partial T > 0$. These reactors with extremely compact cores operating at higher temperatures are prone to disaster; they are not the bases for the 1985 reactors outlined here, but there is a finite chance that the core of the reactors being considered could be rendered more compact in an accident, so that this unstable behavior was realized.

[16] Decreasing the volume of fuel in the core, and increasing the capacity of the sodium coolant and transfer systems so as to maintain sodium outlet temperatures, results in the undesirable larger positive values of $\partial\rho/\partial T$. [Cf. General Electric Rept. GEAP-4418, Section 3.3.2, "Evaluation Techniques for Survey Data and Their Graphic Representation, Survey of Fast Reactor Cores"; cf. also Cohen, and O'Neill, "Reactor Design," Section 3.2.] But this design variant also increases the rate of production of new fissile fuel so as to decrease fuel inventories. There is a gain in expected design productivity, then, from reducing the certainty of this design productivity. The General Electric trade-off appears to be an addition to reactivity = 0.005 for every 1 per cent and an increase in the breeding ratio by as much as 20 per cent.

2. The Gas-Cooled Fast Breeder Reactor

Designs for transferring heat from the reactor are available using helium, carbon dioxide, or sulfur dioxide gas as the transfer medium. One of these shows remarkably promising productivity. The helium gas system is of great interest because it is well advanced toward adoption in a sophisticated thermal reactor demonstration plant.

Helium is introduced into a coolant system from outside of the reactor through filters and compressors, and it enters the reactor container and finally the core by means of large gas compressors at 500°F and 1,050–1,060 psia.[17] The gas is blown up into the fuel where its temperature is increased to 1,150°F and its pressure is reduced approximately to 1,015–1,020 psia, and then is blown down through four steam generators surrounding the core. The blowers, core, and generators are all contained in a single vessel of prestressed concrete or steel and this vessel is contained within a secondary reinforced concrete structure. The helium emerges from the two pressure vessels at 500°F and 1,000 psia for purification and storage, and the steam emerges at 950°–1,000°F and 2,400 psia for transfer to the turbine-generators. The circuit is completed outside of the pressure vessel by condensing the steam and running the condensate through feed pumps for re-entry to the steam generator intake, and by taking the helium from storage to the gas compressors for re-entry as well.

This design shows differences with the liquid sodium reactor layout. The largest components — gas compressors, fuel core, steam generators, primary and secondary pressure vessels — include three that are not of comparable magnitude in the other reactor. The compressors in a 1,000 MW gas reactor are much larger than the pumps in the liquid metal system (and are beyond present experience with all large pumps, for that matter). The steam generators are larger than those found in the liquid metal-cooled reactor. Moreover, the primary and secondary pressure vessels are much larger and more complex, because of the necessity to contain the fission products in high-pressure helium when an accident opens the gas piping or the heat transfer surfaces.

The design differs in other dimensions from those for the liquid metal fast breeders. It is simpler: the four gas-to-steam heat transfer loops in the primary vessel are once-through steam generators, not primary to intermediate loops carrying more coolant to final loops containing steam. System reliability is sought through particular design variants

[17] These values are "representative" design values derived from the two most detailed design studies of a GCFBR: *Gas-Cooled Fast Reactor Concepts*, Oak Ridge National Laboratory, Reactor Division, Rept. ORNL-3642 (September, 1964); and General Atomics Rept. GA-5537.

not found in the liquid sodium reactor. The helium reactor is designed to pour steam and water into the core, if the coolant is voided by accident, to prevent core meltdown.[18] As a result, the safety techniques and routings in the helium design are much more straightforward. Also the containment of compressors, core, and heat transfer surfaces in the pressure vessel in the helium reactor is substantially more complete. It is prompted by the high pressures under which the helium operates, as compared to the ambient pressure of liquid sodium, and it adds to the capital components over and above those for containment in the liquid sodium reactor.

In terms of economic performance, where such performance is measured by $Q_n = \alpha K^{\beta'} F^{\psi'}$, the helium reactor has characteristics that reflect these design differences. The simplicity of the system reduces the number of units of capital K, but the pressure requirements increase the sizes of units of capital, particularly in containment. On the other hand, this gas is a relatively poor heat transfer medium, so that the thermal energy capability of a kilogram of plutonium — that is, the specific power of the fissile fuel — is lower than for sodium reactors at the same turbine throttle temperature and pressure. This requires an increase in the plutonium content in a fuel loading for any given level of megawatt output. The net effects of these design characteristics on the relative requirements of capital and fuel for Q_n can be shown by defining β' and ψ' for a 1975–1980 helium-cooled fast breeder reactor.

Calculations of approximate β' and ψ' cannot be made from a single design study, or even from a group of related studies similar to those by General Electric on the liquid metal fast breeder. Detailed core analyses for gas-cooled reactors have been completed by the Gulf General Atomics Company which provide the parameters for estimating fuel inventories for a wide variety of designs.[19] The study of *Gas-Cooled Fast Reactor Concepts* by Oak Ridge National Laboratory specifies the capital components to go with fuels and different gas coolants for different reactor capabilities.[20] A consistent or single design has to be constructed *de novo* from the characteristics for parts of a helium-cooled 1,000 MW fast reactor shown in these two sources. Then this design can be varied for size and for specific power. First estimates of

[18] The liquid metal reactor cannot rely on this technique because of the volatility of sodium in water, so that elaborate additions to capital are made in the LMFBR to reduce the probability of voiding, and elaborate auxiliary safety rod or sodium storage systems are constructed to cut off reactivity increases if voiding ever does take place. Cf. Cohen and O'Neill, "Reactor Design," and General Atomics Rept. GA-5537.

[19] General Atomic Rept. GA-5537.

[20] Oak Ridge National Laboratory Rept. ORNL-3642.

K_0, F_0, MW_0, and then of the effects on K, F, and MW of scale and of specific power variations, follow from these design variants.

The most important design feature is the integrated primary and secondary containment vessel, with the gas compressors, fuel core, and gas-to-steam heat transfer surfaces inside the primary vessel. The ORNL study shows the costs of these components as more than $37.9(10^6)$ for capability of 1,000 MW. With the addition of helium transfer, purification, oil removal, and low-pressure leak-off recovery systems (all outside of the containment vessels) total capital costs are $52.1(10^6)$ for 85 components. The average size of a unit of capital $\sum PQ/\sum P$ equals $52.1(10^6)/$32.0(10^6)$ or 1.63, so that the total amount of capital is (1.63)(85) or 139 units of roughly equal size. For size comparable to that of a unit in the liquid metal reactor, this is 522 units, because the average price per unit is 3.76 times that in the liquid metal reactor.

The inventory of fuel required to complement these units of capital can be found after specifying the particular uranium–plutonium mixture of interest and the specific power of the mixture. Oxides of plutonium are to be fabricated to perform at specific powers close to 800 kW_t/kg of this fuel; at least fuel requirements are not reduced by higher specific powers, for fabrication requirements increase at such a rate as to compensate for any further decrease in oxide content beyond this level.[21] An initial inventory for 782 kW_t/kg at 1,000 MW is shown as $N_0 =$ 3,350 in the General Atomic analysis. The lifetime inventory is equal to $N_0[1 - (B^{1/t} - 1)] \sum_1^{30} (1 + r)^{-t} = -351$ kg, given that the annual reduction of in-place requirements is 11.7 per cent because of a ratio of fissionable products produced to consumed of 1.44, a core lifetime t of 3.3 years, and an annual use charge of 10 per cent for plutonium. The gain in produced fuel in later years more than compensates for the use charge even in present value terms. The gas breeder produces more fission fuel than it consumes. Then $K_0 = 522$, $F_0 = -351$, and $MW_0 = 1{,}000$ for a design based on the ORNL specifications for a large helium-cooled fast reactor.

Redesigning this reactor to decrease the specific power increases the fuel inventory and reduces the capital requirements. The variant allows

[21] Cf. General Atomic Rept. GA-5537, p. 40, where it is argued that "raising the rate can lead to increased fuel cost if pursued too far. This arises because, although inventory charges are inversely proportional to rating, a component of the fuel fabrication cost increases with rating (because there is a limit to the power available per unit total length of fuel element, set by internal heat conduction, regardless of the coolant median) so somewhere there is an optimum as far as total fuel cost is concerned. The question is dealt with at greater length in Section 12, where it is shown there is little or no economic incentive to use ratings much over some 800 kW/kg." *Ibid.*

an estimate of $\partial K/\partial F$, the trade-off of capital for fuel in the helium-cooled reactor's production function. The increase in the fuel input is 228 kg in a reactor lifetime as the specific power is reduced from 800 kW/kg to 718 kW/kg.[22] The reduction in capital is more difficult to assess. It appears from these studies of design variants that the simplification of the core, and the reduction of helium temperature and pressure, as a result of (782 – 718)/782 per cent power reduction, has the same effect on capital requirements as that per cent reduction in thermal megawatt capacity. The ORNL design variants for the helium-cooled fast reactor show an average reduction of 0.6 per cent in units of capital contained in the primary and secondary pressure vessels for each 1.0 per cent reduction in capacity over the range 480 to 1,000 MW; on this basis, the 522 units of capital are reduced by 24 units, as a result of the 7.8 per cent power reduction. Then, from the reduction in specific power, capital is reduced by 24 units and fuel is increased by 228 units, or $\partial K/\partial F = -11/100$.

Most of the effects of the scale of output are shown by other design variants. A reduction in scale, from 1,000 MW to 525 MW as one variant in the ORNL study, brings capital in the pressure vessels down by approximately 50 per cent. Since the requirements for peripheral equipment — purification and storage components — are not changed substantially, then the number of units of capital is 405 rather than 498 at specific power of 718 kW/kg. The effects on fuel inventory are shown by the fuel cycle design details for the 525 MW helium fast reactor variant. The initial inventory is 2,038 kg for this size of the general design at specific power of 646 kW/kg; with an overall breeding ratio of 1.5 and a core lifetime of 3.6 years, the 30-year requirement of fuel in present value terms is –18 kg. The initial inventory is reduced to an amount close to 1789 kg, and the lifetime inventory to –187 kg, by increasing specific power to 718 kW/kg. Then this smaller reactor utilizes slightly more fuel over the inventory lifetime than does the larger reactor — –187 kg, rather than –351 kg — but as noted in the General Atomic study, "fuel cycle cost [that is, inventory per unit of capacity] does not vary greatly with reactor size, for cores of 5,000 liters or larger."[23]

[22] The fuel cycle analyses show that a reduction from specific power of 782 kW/kg to 718 kW/kg involves a decrease in burn-up from reduced core temperatures sufficient to reduce B from 1.17 to 1.04. This increases the present value of inventory from –351 kg to –123 kg, for a decrease in specific power of 64 kW/kg. It is assumed that there is no further increase in inventory from a reduction in specific power from 800 kg to 782 kg — or that the inventory required at these two specific powers is roughly the same (as in the previous footnote).

[23] General Atomic Rept. GA-5537, p. 162.

These estimates can be used to outline the dimensions of the production function. The trade-off of the two inputs $\partial K/\partial F = -\psi' K_1/\beta' F_1 = -11/100$, which with $K_1 = 498$, $F_1 = -123$, at $MW_1 = 1{,}000$ and specific power of 718 kW/kg is equal to $-\psi'(498)/\beta'(-123) = -11/100$. A second equation follows from the effects of scale; the ratios of the observations of inputs and of outputs in the production function are $\log(1{,}000/525) = \beta' \cdot \log(498/405) + \psi' \cdot \log(-123/-187)$ which reduces to $2{,}954 = 1{,}897\beta' - 1{,}820\psi'$. Solving these two equations for $\beta' = 1.46$ and $\psi' = -0.04$, the first approximation to the production function for the helium-cooled fast breeder reactor appears to be

$$Q_n = \alpha K^{1.46} F^{-0.04}$$

for megawatts of capacity in terms of 30 years of capital and fuel requirements.

3. Safety Characteristics of Gas-Cooled Breeder Reactors

As shown in the case of the liquid metal reactor, the capacity of a fast breeder reactor concept can vary from the level called for in the design, as a result of transient temperature changes in the reactor core. In the liquid metal-cooled reactor designs, both *a priori* and experimental analyses have shown that, at certain very high fuel temperatures, incursions of more temperature or power will add to reactivity so as to lead to further temperature and reactivity increases. A departure from design temperature can lead to disequilibrium operation of the fuel core, with no self-adjustment back to equilibrium. The core can be shut down before it melts as a result of disequilibrium; this shut-down one way or another takes the operating value of MW to zero. But the designs for working reactors avoid as much as possible this extreme result. Rather, the output potential is curtailed in favor of smaller variations, in keeping with the production function $MW = \alpha K^{\beta'} F^{\psi'}$, with $\alpha = \alpha_0 \cdot e^{-(\Delta T/T)(1+\xi)}$ as the productivity effects from any transient temperature T greater than design temperature T^*. A similar characterization of the effects of temperature changes can be made for both the sodium-cooled and gas-cooled reactor designs.

Experiments with reactivity and temperature point to contrasts in the behavior of gas in the fuel core with that of liquid metal. Temperature surges can be expected to occur infrequently in helium during circulation of the coolant; for one thing, the occurrence of gas bubbles, as in liquid sodium, cannot lead to local hot spots which become temperature surges. But coolant temperature will increase much more rapidly when gas pumping systems stop, because the lower heat absorption of this coolant places more reliance on flow to reduce core temperature. Then the way to achieve reliability of gas reactors is to prevent voiding of the coolant from the core.

The adverse results from voiding are greater when the fraction of the core given over to coolant is smaller. The poor results are also more pronounced when the total pumping power required to circulate the coolant through the core is larger. If there is an outage in coolant system pumping, the gas reactor core is subject to the larger temperature increase: more fuel per unit volume and more reliance on pumping to remove heat from this fuel result in more in-core heat in such circumstances.

The temperature change reduces plant capability more in the helium-cooled reactor. For any given number of degrees of temperature change, there is more induced change in reactivity in the helium and plutonium core. The reactivity effect $\partial\rho/\partial T$ is parceled into ρ_1 and ρ_2 in both designs. The value for $\partial\rho_1/\partial T$, the decrease in reactivity ρ from decreased coolant density at higher temperatures, is less in helium because its density does not vary greatly; the value of $\partial\rho_2/\partial T$, where ρ_2 is "degrading of the energy spectrum" so as to increase fission, is positive and large in both systems. Then the overall change in reactivity in the gas reactor is more likely to be positive, since it could be dominated by the value of $\partial\rho_2/\partial T$ over that of $\partial\rho_1/\partial T$ in this case. The design has to hold back this reactor to a greater extent — in other words, the gas reactor must be shut down more rapidly so that runaway expansive reactivity-to-temperature feedbacks do not take place.[24]

The trade-offs over the range of present values of specific power for temperature and pressure safety are likely to favor liquid metal over gas. The temperature coefficients of reactivity for gas-cooled reactor designs

[24] Substantial temperature changes from complete loss of pumping power, and thus from voiding of coolant in the core, are usually prevented from going too high by shutting down of the reactor. There has to be some way devised, in fact, for shutting down quickly enough to prevent melting of the core and, at the limit, causing a nuclear explosion. Present plans in gas reactors center on emergency routings which flood the core with water from the steam systems, which will reduce temperature, but, by greatly reducing neutron leakage out of the core, will also increase reactivity. For safety — or reliability of thermal megawatts of capacity — there must be absorption of the additional neutrons, presumably by using core additives or "poisons" such as boron. But these additives affect the design level for production of thermal energy, so that a trade-off — yet to be specified — has to be made of design thermal power for decreased probability of zero power for extended periods in the reactor lifetime. Cf. J. W. Hallam, R. K. Haling, P. Killian, and G. T. Peterson, "The Flood Safety of Steam- and Gas-Cooled Reactors" (U.S. Atomic Energy Commission, Contract AT(04-3)180, PA 13, 1965), Figure 1; cf. also, G. Sofer *et al.*, "Conceptual Design and Economic Evaluation of a Steam-cooled Fast Breeder" (United Nuclear Corporation Rept. under contract AT 30-1-20303 (XII), USAEC, 1961), pp. 79-82; and General Atomic Rept. GA-5537, p. 72 *et seq.*

are positive under a wide range of conditions because of the lack of a "canceling effect" from substantial resonance absorption of neutrons at higher temperatures (here shown as $\partial\rho_1/\rho T$, the "Doppler effect").[25] Then the necessary shutdown time for the first reactor is likely to be less, either because of less chance of meltdown or of having to close down the reactor at $T > T^*$ to prevent such core meltdown. This is shown by considering in $Q_n = \alpha K^{\beta'} F^{\psi'}$ that $\alpha = \alpha_0 e^{+x}$ with $x = -(\Delta T/T)(1 + \xi)$, the product of the temperature variant and one plus the elasticity of power with respect to temperature, as in the liquid metal production function. The lowest values of ξ in the range 480 to 1,000 MW are for the steam-cooled and liquid metal breeders with the gas-cooled reactor reaching much higher values. To scale capacity in terms of relative reliability, $\xi = -0.008$ for the gas reactor. Then the production function of the gas-cooled fast breeder is

$$Q_n = (0.067)K^{1.46}F^{-0.04}e^{(-\Delta T/T)(1-0.008)}$$

The two types of fast reactors require different amounts of both K, F for given megawatts of capacity, and with differing degrees of reliability.

[25] Cf. M. Dalle-Donne, "Comparison of Helium, Carbon Dioxide, and Steam as Coolants of a 1,000 Megawatt Electric Fast Reactor," Euratom Gesellschaft für Kernforschung mbH, 1965, Table 4 for Doppler coefficients and Cohen and O'Neil, "Reactor Design," p. 91 and Table 2 for Doppler coefficients for the liquid metal fast breeder reactor shown in the General Electric design study.

Appendix E

The Computer Model of Energy Production Without Fast Breeder Reactors

R. L. Schmalensee and P. W. MacAvoy

Notation

We have four size classes, denoted by j = 1, . . ., 4; and nine regions, denoted by i = 1, . . ., 9. The principal driving variables are the following:

QTi	Total demand for generating capacity in Region i
PINK	Productivity increase in nuclear capital
PIFK	Productivity increase in fossil capital
PIFF	Productivity increase in fossil fuel
TE	Thermal efficiency of thermal reactors
RT	Fuel loading of thermal reactors — tons of U_3O_8 over a 30-year lifetime
WjT	The fraction of total demand for plants in size class j

The QT's were calculated by extrapolating historical relationships. All the others are entered in TROLL as parameters: they can be changed from run to run. In previous runs we have set TE = 0.4. (RT should generally be about 9.0 and TE should be between 0.3 and 0.4.)

The historical size distribution of plants gives W1T = 0.30, W2T = 0.15, W3T = 0.10, and W4T = 0.45. The PI's give the ratio of 1970 costs to 1985–2005 costs in each category. Their value depends on the rate of improvement per year and the number of years involved. A table may clarify things:

Values of the PI's

Progress per Year	Years 15	20	25	30	35
0.5%	1.0777	1.1049	1.1328	1.1614	1.1907
1.0%	1.1610	1.2202	1.2824	1.3478	1.4166
1.5%	1.2502	1.3468	1.4509	1.5631	1.6839
2.0%	1.3459	1.4859	1.6406	1.8114	1.9999

The 1970 costs are built into the equations: they will be given below.

We will examine first the market for reactors, then the fuel market, and finally the consumers' surplus calculations.

Reactor Market

The principal quantities determined in this section are:

- QNTO Total MW_e of thermal reactor capacity operating
- QNTT MW_e of thermal reactor capacity installed in the current five-year period
- QNTi MW_e of thermal reactor capacity installed in region i
- Sij The market share of thermal reactors in size class j in region i
- Si The overall market share of thermal reactors in region i
- S The overall market share of thermal reactors

We have a number of identities connecting these quantities:

$$QNTO = QNTO(-1) + QNTT \quad (1)$$

$$QNTT = \sum_{i=1}^{9} QNTi \quad (2)$$

$$QNTi = Si*QTi \quad \text{(9 equations)} \quad (3)$$

$$Si = \sum_{j=1}^{4} Sij*WjT \quad \text{(9 equations)} \quad (4)$$

$$S = QNTT/ \sum_{i=1}^{9} QTi \quad (5)$$

Total capacity operating in 1985 is taken as 100,000 MW_e The Sij are determined by the 36 share equations:

$$Sij = \exp(-4.375 + .418*\log(TE*ASj) - .859*\log(PNFT/(PFFi/PIFF)) - 1.030*\log((PNKj/PINK)/(PFKj/PIFK))). \quad (6)$$

The variables ASj, PFFi, PNKj, and PFKj do not appear in any equations as such; they are replaced by their numerical values. They are, respectively, the average size of a plant in size class j in thermal megawatts, the 1970 fuel cycle cost of fossil-fueled plants (mills per kilowatt hour), and the 1970 costs per MW_e of nuclear and fossil capital in any size class j. The values to be used in the equations are given in the tables below.

Region	PFF
1	3.07
2	2.36
3	1.85
4	2.35
5	1.77
6	1.68
7	1.88
8	1.97
9	3.00

Size Class	AS	PNK	PFK
1	234	360	290
2	625	340	244
3	1250	157	153
4	2500	130	133

Two other quantities of interest are calculated by this block of equations:

Sj The market share of thermal reactors in size class j
PNKTB The average price per MW_e of thermal reactors sold

Both are determined by identities:

$$Sj = \sum_{i=1}^{9} Sij*QTi / \sum_{i=1}^{9} QTi \qquad \text{(4 equations)} \qquad (7)$$

$$PNKTB = [\sum_{j=1}^{4} WjT*Sj*PNKj / \sum_{j=1}^{4} Sj*WjT] / PINK \qquad (8)$$

Uranium Market

In this section, we determine the following:

PNFT The fuel cycle cost of thermal reactors — mills per kilowatt hour
PU The price of uranium oxide — dollars per ton
DUT New demand for uranium oxide — tons of reserves
TDUT Total demand for uranium oxide — tons of reserves
SUT Supply of uranium oxide — tons of reserves

As before, there are identities:

$$TDUT = DUT + TDUT(-1) \tag{9}$$

$$SUT = TDUT \tag{10}$$

The fuel cycle cost is calculated by a relation derived from a diagram in *The Westinghouse Engineer*[1]:

$$PNFT = .75 + .058*PU \tag{11}$$

The supply curve was fitted from a table in *The 1967 Supplement to the 1962 Report to the President on Civilian Nuclear Power*. It is normalized on PU:

$$PU = (\log(SUT) - 11.3) / .16 \tag{12}$$

The demand curve assumes that reserves must be on hand in an amount equal to the present value of all future loadings of installed capacity. We assume that a reactor requires R kg of uranium fluoride per megawatt per year for 30 years, and we use a discount rate of 10 per cent. Since 1 kg = 2.2 lb, and 1 lb uranium fluoride requires 3.5 lb uranium oxide, we have 7.7R lb uranium oxide per megawatt per year. Discounting over 30 years at 10 per cent, we get a present value of 80.08R lb uranium oxide per megawatt or 0.0404R tons. We thus have

$$DUT = .0404*R*QNTT \tag{13}$$

TDUT in 1985 is taken as 4.0(100,000) = 400,000.

Consumers' Surplus

If all the PNKj's were increased by 1 per cent, all the Sij's would fall by 1.03 per cent from the share equations. Ignoring the feedback via the fuel markets, we can thus take the partial equilibrium elasticity of demand as −1.03. (This will be an overestimate of the general equilibrium elasticity as a lower share tends to lower fuel prices and increase

[1] J. C. Rengel "High-gain breeders can stabilize uranium fuel requirements," *The Westinghouse Engineer*, Vol. 28, No. 1 (January, 1968), pp. 3–7.

demand. Since the supply curve is very elastic, however, this is probably not a bad approximation.)

If $Q = A*P^{-1.03}$, we can calculate

$$AT = QNTT*PNKTB^{1.03} \tag{14}$$

The area under this curve from 0 to QNTT is given by

$$\frac{AT^{.97}QNTT^{.03}}{.03}$$

Subtracting producers' revenues we obtain consumers' surplus:

$$CST = QNTT * PNKTB / .03 - QNTT * PNKTB$$

Appendix F

The Computer Model of Fast Breeder Reactor Demands and Prices

R. L. Schmalensee and P. W. MacAvoy

Notation

Again we have four size classes, denoted by j = 1, . . ., 4, and nine regions, denoted by i = 1, . . ., 9. This appendix will deal with the determination of equilibrium quantities in one five-year period. Questions of the length of the period to be considered will not be taken up explicitly here. The following driving variables are in common with the nonbreeder world (NBW):

QTi	Total demand for generating capacity in region i
PINK	Productivity increase in thermal capital
PIFF	Productivity increase in fossil fuel
TE	Thermal efficiency of thermal reactors
WjT	The fraction of total demand that is for plants in size class j

The QTi's were calculated by extrapolating historical relationships. They will cover the period 1985–2004 in four five-year periods. All the others are entered in TROLL as parameters; they can be changed from run to run.

The following driving variables pertain only to th. breeder world (BW):

PPR	The rate of plutonium production from thermal reactors — kilograms per megawatt per year (80 per cent load factor)
BE	The thermal efficiency of breeder reactors
RB	The fuel loading of thermal reactors, tons of uranium oxide per megawatt over a 30-year life
PPB	Plutonium price — dollars per gram
DG	Dummy variable: one if gas-cooled breeders are present, zero otherwise
DL	Dummy variable: one if liquid metal breeders are present, zero otherwise

N The number of firms producing fast breeder reactors

GL Kilograms of plutonium required per MW_e of gas-cooled breeder reactor capacity over a 30-year lifetime

LL Kilograms of plutonium required per MW_e of liquid metal breeder reactor capacity over a 30-year lifetime

PPB, DG, and DL will be exogenous data files, with one entry for each period. The D's enable us to take into account different dates of introduction. All the other parameters hold for all periods and can be varied from run to run.

PPR will vary between 0.2 and 0.4, BE will probably be around 0.4, and RB will be greater than or equal to RT, since there will be no recycling in a world with breeders. PPB will initially take on a value of $10 for all periods.

The reason for this is that it is unlikely that the government will dump its plutonium inventory on the open market when breeders are first introduced. It is unrealistic, therefore, to let the market clear in the early years of breeders: if this were done, the price would fall very low and then rise rapidly once the existing stocks were all committed. We will instead fix the plutonium price outside the model and keep track of excess demand within the system. If there proves to be excess demand at a price of $10 in some periods, we can raise the price in later runs until the excess is eliminated. The assumption here is that price will probably not fall in response to excess supply, but that it will rise when the government faces excess demand.

The L's are obtained from the design studies in the text. We first find the lifetime inventory per MW_e for the two design studies of the two reactor types. We then weight these by the fraction of demand in the size class they come from. Since we have studies of reactors in only the two largest size classes, this procedure has the effect of ignoring small reactors. The work involved may be most clearly shown in table form:

	Liquid Metal		Gas-Cooled	
	large	small	large	small
MW_t	2500	1395	2600	1317
$MW_e(0.4\ MW_t)$	1000	568	1040	527
Fuel (kg)	558	729	−351	−187
kg/MW_e	0.558	1.283	−0.337	−0.355
Weight	0.45	0.10	0.45	0.10
Average	0.689(LL)		−0.341(GL)	

The rest of this appendix will not follow the outline of the preceding appendix. Rather, we will first discuss the uranium and plutonium markets. The marginal cost and price equations for breeders are next, followed by the reactor market. Finally, we will deal with consumers' surplus — where the breeder world and nonbreeder world are compared.

Uranium Market

In this section we determine the following:

PTF	The fuel cycle cost of thermal reactors — mills per kilowatt hour
PUB	The price of uranium oxide — dollars per ton
DUB	New demand for uranium oxide — tons of reserves
TDUB	Total demand for uranium oxide — tons of reserves
SUB	Supply of uranium oxide — tons of reserves

The following equations are the same as in the NBW — except, of course, for the different variables appearing in them:

$$\text{TDUB} = \text{DUB} + \text{TDUB}(-1) \qquad (1)$$

$$\text{SUB} = \text{TDUB} \qquad (2)$$

$$\text{PUB} = (\log(\text{SUB}) - 11.3)/.16 \qquad (3)$$

$$\text{DUB} = \text{RB}*\text{QBTT} \qquad (4)$$

As before, we take TDUB in 1985 to be 400,000. (QBTT is newly installed thermal reactor capacity; it is determined in the reactor market.)

To determine the fuel cycle cost of thermal reactors, we must add to Equation 11 in the preceding appendix a term reflecting revenues from the production of plutonium. The graph in *The Westinghouse Engineer* from which Equation 11 was derived suggests that this term should be on the order of $-0.05*$PPB.

We can check this and allow for differences in PPR by a more circuitous route. PPR is kg/MW_e of capacity/year, assuming an 80 per cent load factor, or (PPR/0.8)kg/MW_e produced/year. There are approximately 8,760 hours per year, so this gives (PPR/7,008)kg/MW_e hours or (PPR/7,008) grams/kilowatt hour. To convert PPB to mills, we must multiply by 1,000, giving our version of this term as $-$(PPR/7.008)$*$PPB. Our result corresponds to the Westinghouse term if a PPR of 0.35 is assumed, and this figure is within the range mentioned above. Therefore, we have the equation

$$\text{PTF} = .75 + .058*\text{PUB} - (\text{PPR}*\text{PPB}/7.008)* \text{(IF DG + DL GRTO THEN 1 ELSE 0)} \qquad (5)$$

Plutonium Market

This section of the model determines the following quantities:

TESP	Total excess supply of plutonium — kilograms
NDP	New demand for plutonium — kilograms
NSP	New supply of plutonium — kilograms
PLF	Fuel cycle cost of liquid metal reactors — mills per kilowatt hour
PGF	Fuel cycle cost of gas-cooled reactors — mills per kilowatt hour

The excess supply equations are mostly bookkeeping: as mentioned above, the price of plutonium will be set so as to keep TESP positive. The equations are

$$TESP = TESP(-1) + NSP - NDP \qquad (6)$$

$$NDP = LL*QBLT + GL*QBGT \qquad (7)$$

$$NSP = (9.891)(PPR)(QBTT) \qquad (8)$$

QBLT and QBGT are the megawatts of liquid metal and gas-cooled capacity installed in the period; they are determined in the reactor market. QBTT is the amount of new thermal capacity operating; it is determined in the reactor market also. The constant term is there because PPR is an annual rate of production, and this then converts an annual rate to a lifetime stock from new thermal reactors. The text gives a value for TESP at the start of the period (1985) of 150 tons, or 150,000 kg.

The fuel cycle cost equations are derived from a graph in *The Westinghouse Engineer*[1]:

$$PLF = .10 + .07PPB \qquad (9)$$

$$PGF = .40 + .005PPB \qquad (10)$$

Breeder Capital Prices

We must first add fabrication costs to PPB and multiply by 1,000 to get the price of fuel per kilogram:

$$BFPK = 1000*(PPB + 11.2) \qquad (11)$$

The source for this is the discussion in the text of plutonium fuel production and price.

[1] See footnote 1, Appendix E.

The following are the important quantities generated by this section of the model:

- PGKj — Price of gas-cooled capital in size class j — dollars per megawatt electric power
- MCGj — Marginal cost of gas-cooled capital — dollars per megawatt electric power
- PLKj — Price of liquid metal capital — in size class j — dollars per megawatt electric power
- MCLj — Marginal cost of liquid metal capital — dollars per megawatt electric power
- DDLj — Dummy variable: one if liquid metal is cheaper in class j, zero if gas is
- DDGj — Dummy variable: one if gas is cheaper in class j, zero if liquid metal is

The first equations relate the marginal cost and price determination for each of the two types for each of the four size classes:

PGKj = (20 − DG*19)*(−1.03*N/ (1 − 1.03*N))*(MCGj) (4 equations) (12)

PLKj = (20 − DL*19)*(−1.03*N/ (1 − 1.03*N))*(MCLj) (4 equations) (13)

Notice that the prices are multiplied by 20 if the system is not present. The reason for this will (hopefully) become clear shortly. The markup or second term is the standard Cournot mechanism. The third term is rather messy.

MCtj = (a** (− 1/c))*(ASj** (1 − c)/c)
*[BFPK ((BFPK*b/90000*d) ** − b/c)
+ 90000* ((BFPK* b/90000*d) **d/c)] / BE (8 equations) (13a)

The source is in the main text. The ASj's are given in Appendix E. When t = L, we are computing the MCIj's and we have

$$a = 6.2*10^{-9} \qquad c = 4.0$$
$$b = 2.9 \qquad d = 1.1$$

When t = G, we are computing the MCGj's and we have

$$a = 0.169 \qquad c = 1.42$$
$$b = 1.46 \qquad d = -0.04$$

The assumption we make is that when both breeder systems are present, the cheaper one will always be chosen. When only one is present, the first term in Equations 12 and 13 will assure that it will be chosen over the absent one. Finally, if neither is present, the choice is irrelevant, as we will see below.

How do we decide which breeder is cheaper? We must include both fuel cycle cost and capital cost, clearly. If we assume an 80 per cent load factor, each MW_e of capacity will result in 7,008,000 kWh each year. Since the fuel cycle cost is in mills per kilowatt hour, 7,008 times the fuel cycle cost (PLF or PGF) gives the fuel cost per MW_e of capacity per year. Discounting this cost over 30 years at a 10 per cent rate of interest, we get 65,665 times PLF or PGF as the lifetime fuel cost per MW_e of capacity. With this in mind, we use the following equations to generate the dummies:

DDLj = (IF 65,665*PLF + PLKj LES 65,665*PGF + PGKj THEN 1 ELSE 0) (4 equations) (14)

DDGj = 1 − DDLj (15)

Reactor Market

The equation that does most of the work here is the share equation. To simplify presentation, we will write here the general form of the share equation:

$$S(A,B,C,D,E) = \exp[-4.375 + .418*\log(TE*A) - .859*\log(B/C) - 1.03*\log(D/E)]$$

The principal quantities determined here are the following:

QBTT Newly installed thermal reactor capacity
QBTO Total thermal reactor capacity
QBLT New liquid metal capacity
QBLO Total liquid metal capacity
QBGT New gas-cooled capacity
QBGO Total gas-cooled capacity
QBkTi New capacity of type k installed in region i
[k = T (thermal), L (liquid metal), and G (gas-cooled)]

(all these figures are in MW_e, as usual)

BSij Breeder share: size class j, region i
GSij Gas-cooled share: size class j, region i
LSij Liquid metal share: size class j, region i
TSij Thermal share: size class j, region i
NSij Nuclear (all) share: size class j, region i

We also deal with some weighted averages of these S's. For instance, BSj is the breeder share in size class j, and LSi is the liquid metal share in region i, and NS is the total nuclear share.

There are a (large) number of identities connecting these quantities:

$$QBTO = QBTO(-1) + QBTT \qquad (16)$$

$$QBLO = QBLO(-1) + QBLT \qquad (17)$$

$$QBGO = QBGO(-1) + QBGT \qquad (18)$$

$$\begin{aligned} GSij &= DDGj * BSij \\ LSij &= DDLj * BSij \end{aligned} \qquad \text{(72 equations)} \qquad (19)$$

$$TSij = NSij - BSij \qquad \text{(36 equations)} \qquad (20)$$

$$kSi = \sum_{j=1}^{4} kSij * WjT \qquad k = B,G,L,T,N \qquad \text{(45 equations)} \qquad (21)$$

$$kSj = \sum_{i=1}^{9} kSij * QTi \Big/ \sum_{i=1}^{4} QTi \qquad k = B,G,L,T,N \qquad \text{(20 equations)} \qquad (22)$$

$$kS = \sum_{j=1}^{4} kSj * WjT \Big/ \sum_{j=1}^{4} WjT \qquad k = B,G,L,T,N \qquad \text{(5 equations)} \qquad (23)$$

$$QBTT = \sum_{i=1}^{9} QBTTi \qquad (24)$$

$$QBGT = \sum_{i=1}^{9} QBGTi \qquad (25)$$

$$QBLT = \sum_{i=1}^{9} QBLTi \qquad (26)$$

$$QBLTi = LSi * QTi \qquad \text{(9 equations)} \qquad (27)$$

$$QBGTi = GSi * QTi \qquad \text{(9 equations)} \qquad (28)$$

$$QBTTi = TSi * QTi \qquad \text{(9 equations)} \qquad (29)$$

Now, all we need are the NSij's and the BSij's. These are computed by the following set of equations:

$$\begin{aligned} SBFij = S(&ASj, DDLj * PLF + DDGj * PGF, PFFi/PIFF, \\ &DDLj * PLKj + DDGj * PGKj, PFKj/PIFK) \end{aligned} \qquad \text{(36 equations)} \qquad (30)$$

$$STFij = S(ASj, PTF, PFFi/PIFF, PNKj/PINK, PFKj/PIFK) \qquad \text{(36 equations)} \qquad (31)$$

$$NSij = (\text{IF } SBFij \text{ GRT } STFij \text{ THEN } SBFij \text{ ELSE } STFij) \qquad \text{(36 equations)} \qquad (32)$$

$$\begin{aligned} SBTij = S(&ASj, DDLj * PLF + DDGj * PGF, PTF, \\ &DDLj * PLKj + DDGj * PGKj, PNKj/PINK) \end{aligned} \qquad \text{(36 equations)} \qquad (33)$$

$$STBij = S(ASj, PTF, DDLj * PLF + DDGj * PGF, PNKj/PINK, DDLj * PLKj + DDGj * PGKj) \quad \text{(36 equations)} \quad (34)$$

$$BSij = (\text{IF } SBFij \text{ GRT } STFij \text{ THEN } NSij * (1 - STBij) \text{ ELSE } SBTij * NSij) \quad \text{(36 equations)} \quad (35)$$

The values to be used for the ASj, the PNKj, and the PFKj are given in Appendix E. The logic behind all this is that the share equation expresses utilities behavior in choosing between any two systems that embody different technologies. Equation 35 is in terms of a dominant nuclear technology and an intruding nuclear technology.

Consumers' Surplus

We compare the breeder and nonbreeder world here. The principal outputs are

MCBB	Average marginal cost of breeders sold
PNKBB	Average price of all nuclear capital sold
QBWT	Total sales of all nuclear capital
CSBA	First consumers' surplus area
CSBB	Second consumers' surplus area
CSBC	Third consumers' surplus area
CSBT	Total gain in consumers' surplus

The first three quantities are computed by the following identities:

$$MCBB = \sum_{j=1}^{4} WjT * (GSj * MCGj + LSj * MCLj) \Big/ \sum_{j=1}^{4} WjT * BSj \quad (36)$$

$$PNKBB = \sum_{j=1}^{4} WjT * (GSj * PGKj + LSj * PLKj + TSj * PNKj/PINK) \Big/ \sum_{j=1}^{4} WjT * NSj \quad (37)$$

$$QBWT = QBTT + QBLT + QBGT \quad (38)$$

We then generate an intermediate quantity analogous to AT in the NBW:

$$AB = QBWT * PNKBB**1.03 \quad (39)$$

The quantities AT, CST, QNTT, and PNKTB are generated in the nonbreeder section of the model. The following equations determine the CSB's:

$$CSBT = CSBA + CSBB + CSBC \quad (40)$$

$$CSBC = (PNKBB - MCBB) * (QBWT - QNTT) \quad (41)$$

$$CSBA = (QNTT**.0291)\ (AB**.9709 - AT**.9709)\ /.0291 \quad (42)$$

$$CSBB = (AB**.9709) * (QBWT**.0291 - QNTT**.0291)\ /\ .0291 - PNKBB * (QBWT - QNTT) \quad (43)$$

Figure F.1 should make clear what the various areas refer to:

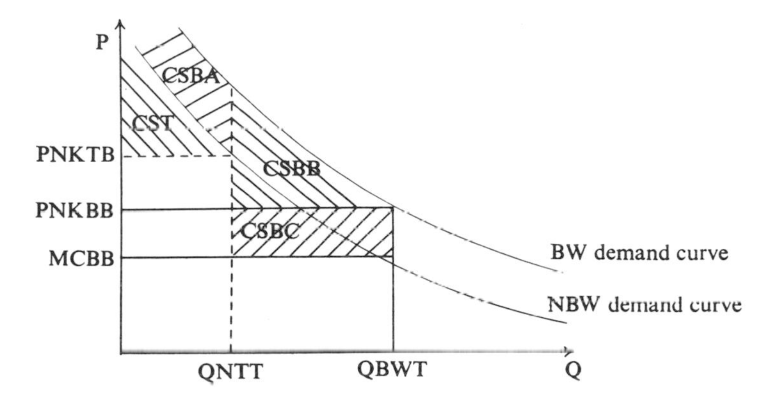

Index